Public Health Ethics Analysis

Volume 6

Series Editor

Michael J. Selgelid,
Centre for Human Bioethics, Monash University, Melbourne, VIC, Australia

During the 21st Century, Public Health Ethics has become one of the fastest growing subdisciplines of bioethics. This is the first Book Series dedicated to the topic of Public Health Ethics. It aims to fill a gap in the existing literature by providing thoroughgoing, book-length treatment of the most important topics in Public Health Ethics—which have otherwise, for the most part, only been partially and/or sporadically addressed in journal articles, book chapters, or sections of volumes concerned with Public Health Ethics. Books in the series will include coverage of central topics in Public Health Ethics from a plurality of disciplinary perspectives including: philosophy (e.g., both ethics and philosophy of science), political science, history, economics, sociology, anthropology, demographics, law, human rights, epidemiology, and other public health sciences. Blending analytically rigorous and empirically informed analyses, the series will address ethical issues associated with the concepts, goals, and methods of public health; individual (e.g., ordinary citizens' and public health workers') decision making and behaviour; and public policy. Inter alia, volumes in the series will be dedicated to topics including: health promotion; disease prevention; paternalism and coercive measures; infectious disease; chronic disease; obesity; smoking and tobacco control; genetics; the environment; public communication/trust; social determinants of health; human rights; and justice. A primary priority is to produce volumes on hitherto neglected topics such as ethical issues associated with public health research and surveillance; vaccination; tuberculosis; malaria; diarrheal disease; lower respiratory infections; drug resistance; chronic disease in developing countries; emergencies/disasters (including bioterrorism); and public health implications of climate change.

More information about this series at http://www.springer.com/series/10067

John G. Francis • Leslie P. Francis

Sustaining Surveillance: The Importance of Information for Public Health

John G. Francis
University of Utah
Salt Lake City, UT, USA

Leslie P. Francis
University of Utah
Salt Lake City, UT, USA

ISSN 2211-6680   ISSN 2211-6699   (electronic)
Public Health Ethics Analysis
ISBN 978-3-030-63926-6   ISBN 978-3-030-63928-0   (eBook)
https://doi.org/10.1007/978-3-030-63928-0

This Springer imprint is published by the registered company Springer Nature Switzerland AG
The registered company address is: Gewerbestrasse 11, 6330 Cham, Switzerland

For our grandson Jack, our granddaughter
Minnow, and our granddaughter Elka
may they live in a healthier and safer world
and
for each other
where it all began

Contents

Chapter 1
Introduction: Why Surveillance Matters

1.1 COVID-19 Stuns the World

In the winter and early spring of 2020, COVID-19 stunned the world. It emerged first in December 2019, in Wuhan, the capital city of Hubei province in central China. As people rushed home for the annual Lunar New Year Festival in late January, the disease spread quickly and widely away from the city of its apparent origins. Shortly thereafter, outbreaks erupted across the globe. A Champions League football (soccer) match held in Milan, Italy, in mid-February turned triumph to tragedy, infecting thousands in northern Italy. The Life Care Center, a nursing home in Washington State, was just the first of many senior care centers in the United States and across the world hard hit by the virus. Within a short six months, over seven million people had been diagnosed with the disease, and just under 6% of them had died.

How could such a public health disaster have unfolded? The world was ripe for a pandemic maelstrom, in many ways. Just over 100 years had passed since the last great pandemic, the 1918–1919 influenza pandemic at the end of the first World War (Barry 2004; Honigsbaum 2019). Other outbreaks of new diseases had come and apparently gone, or at least not travelled very far: SARS, avian influenza, H1N1 influenza, and Ebola. Smallpox had been eradicated and polio brought under control in many areas by vaccination. Even HIV, despite its trail of death and misery, had largely been tamed to a chronic disease, at least where treatment was available. As infectious disease lost salience, so did surveillance.

At the same time, forceful rapids were eroding the capacities of public health. Inequality and unrest were rising. Responses to refugee crises drew the attention of international organizations. Political polarization was growing along with mistrust of government and science, especially in the United States but elsewhere as well. The Trump Administration increasingly threatened or withdrew the United States from international organizations. The World Health Organization's credibility had

© Springer Nature Switzerland AG 2021
J. G. Francis, L. P. Francis, *Sustaining Surveillance: The Importance of Information for Public Health*, Public Health Ethics Analysis 6,
https://doi.org/10.1007/978-3-030-63928-0_1

lagged due to questionable handling of the H1N1 influenza pandemic in 2009 and the Ebola outbreak of 2014–2016 in West Africa. Although the International Health Regulations had entered into force in 2007, fewer than a quarter of states parties had implemented required core capacities by the target date of five years later (Suthar et al. 2018).

Inadequate attention to surveillance was part of the problem. Initial reports from China downplayed risks of human-to-human transmission of the novel coronavirus that was causing respiratory infections in Wuhan. The actual extent to which China improperly suppressed information remains contested. As international travel continued and cases were identified in Europe and the United States in late January, the WHO still judged that it lacked sufficient information to declare a Public Health Emergency of International Concern until January 30, 2020 (WHO 2020). Testing was insufficiently available to detect community spread of the disease even as late as six months into the pandemic. Leaders of some governments, particularly President Trump in the United States, downplayed the importance of surveillance and the critical need to obtain and use the information it provided. Some governments actively suppressed accurate information about the spread of the pandemic. Right-wing groups used social media to circulate misinformation and stoke mistrust of public health measures, even to the point of asserting COVID-19 to be a hoax (Bridgman et al. 2020; Jamieson and Albarracin 2020; Simonov et al. 2020). In short, surveillance was inadequately conducted, communicated, trusted, and acted upon.

We will never know how the COVID-19 pandemic would have unfolded with better surveillance for public health. What we do know is that the pandemic brought widespread death, disease, and economic loss. If surveillance had helped to mitigate even some of this pain, its utility would have been unquestionable.

Our goal in this volume is to present an account of the ethics and politics of surveillance for public health. Pandemics of the level of COVID-19 or the Great Influenza of 1918–1919 may strike only once every hundred years but the critical need is to be ready for them. Our hope is that our work can serve as a basis for warranted trust in surveillance, explaining both its importance and ethical limits. As COVID-19 so tragically illustrates, the world needs sustainable surveillance now more than ever.

1.2   The Ubiquity of Surveillance

Surveillance is often called the "eyes" of public health. With today's technologies, it is fair to say that surveillance may also be the ears, nose, tongue, and even touch of public health. Surveillance has evolved far beyond counting the population and finding cases of dangerous disease to include an almost unimaginable range of ways to understand and address the health of individuals, groups, and populations.

This book is about the ethics and politics of public health surveillance. It places ethical questions about surveillance in the context of political theory, public policy, and the law. The authors are a political scientist and a professor of law and philosophy with specialties in comparative politics, comparative public policy, European politics, regulation, health law, disability law, and bioethics. We do not claim to produce a comprehensive theory of surveillance; the practice is too protean. Rather, we consider the most common surveillance practices by and for public health, how they developed and are continuing to evolve, their benefits, risks, and the ethical challenges they raise. Overall, the goal is to consider whether, why, and how, public health surveillance can be conducted ethically and sustainably. For public health surveillance also encounters resistance, from armed rebels attacking health care workers, to people hiding from contact tracers, to political actors concerned to keep public fears at bay, to privacy advocates seeking destruction of data collected or used without permission. Sometimes these responses succeed and sometimes they are fully justified, sometimes they are both, sometimes they are neither and sometimes they cause grievous harm. Continued and effective surveillance requires warranted trust on the part of those under surveillance and trustworthiness on the part of those conducting the surveillance. Yet when surveillance for public health may be readily linked to surveillance for national security or to surveillance for moral condemnation, trust may no longer be warranted or achieved. Discussions of the ethics of public health surveillance are not a subject of ideal political theory; this is a book about pressing ethical problems for an imperfect, often unjust, and risky world.

Surveillance is ubiquitous. Across the globe, closed circuit cameras stare down from nearly every corner, license plate capture cameras record vehicles passing through intersections, and cell phone towers triangulate users' locations with remarkable precision. Monitors of air and water quality may be linked to geolocation data collected by personal cellphones to reveal exposure risks. Medications may come with sensors that enable prescribers to monitor whether they have been ingested by patients. Internet sites trace users' browsing habits over time and social media sites analyze interconnections among users. Restaurant reservations, ticket purchases, and product purchases can be recorded, saved, analyzed, and repurposed as well. Online book merchants can determine what people start reading, what they emphasize with underlining, how long they linger on particular passages, and when they abandon a manuscript. Predictive analytics generate suggestions about where people may be and what they may be doing—and may profile whether their activities pose risks to themselves or others. Neither time nor distance limits the potential for individuals to be tracked and traced in myriad ways.

Quite complete portraits of individuals' lives can be assembled from the information available in electronic form today. Features of their own health, the health of those around them, and the health effects of their social circumstances are particularly useful pieces of these portraits. Individuals' geolocations reveal visits to health care providers from abortion clinics to dialysis facilities and their web searches indicate interests in health information and products. Changes in patterns of behavior—such as ceasing to go out to dinner or drive to the grocery store—may reveal changes in health status. Information given out by individuals may contain or allow direct inferences about the health of those around them, as when new parents join social media support groups for children with rare genetic diseases or adults in midlife search reports about the quality of nearby dementia care facilities. Predictive analytics may yield inferences about individuals even when no data from them have been used to generate the algorithms imputing characteristics to them. Online medical records may be part of the mix, too, if they are not protected by effective security and privacy rules, or if they are made available in supposedly de-identified form; these records contain far more than medical information and are particularly valuable for identity thieves.

Social media with its global ubiquity offers unparalleled opportunities to collect data at a high level of granularity. At its best, social media empowers people to gather much needed information and connect with others with similar interests and concerns. At its worst, it can spread fear, confusion, or complacency. Surveillance thus goes beyond assembling portraits of individuals to identifying, fostering, or even creating networks. Networks may be made up of people with direct connections to one another. Or, they may include people with shared characteristics who have never encountered one another, even indirectly. Surveillance has the potential to identify everyone with the same rare gene variant, the same common ancestor, the same unusual dietary preferences, the same opposition to vaccination, or the same political views. With such surveillance, networks can be served information, misinformation, alerts, or calls to action. Networks can also be given information about how to communicate with one another or meet one another at a common time and physical place or virtual space. Networks can crowd-source, bringing people together to solve problems, attack outsiders, or create chaos. And, networks can be used to assess what are perceived to be political threats, from efforts at regime change to bioterrorism. Information gleaned from surveillance enhances the power of social media for better or worse, in ways we are only beginning to understand.

Moreover, information gained through surveillance is not static. Surveillance information can be fed into learning systems, such as learning health care systems or learning public health systems. These systems, continually updated in real time, can generate recommendations based on algorithms that in turn are continually learning about their predictive success or failure. These recommendations may include where to deploy surveillance. Exploring the potential of such "artificial" intelligence (AI) for health care is only in the beginning stages, but it is already clear that information gleaned from surveillance is at the heart of these efforts. It is also clear that the algorithms used in AI may be biased in ways that are unjust or damaging to some and that advantage others.

Growing unease attends such ubiquitous surveillance. Responses to its benefits, power, and threats have varied. The European Union General Data Protection Regulation (EUR-Lex 2016) is perhaps the most comprehensive and protective legal structure, allowing a right to be delisted by search engines and requiring real-time consent to some uses of specified forms of identifiable sensitive information, including health information, albeit with an exception for information necessary for public health. Even in the US, with its far more freewheeling approach to information distribution and use, the Supreme Court has held that police must have a warrant to justify searches of cell phone location data collected over a month-long period of time (*Carpenter* 2018). The era in which an executive of a (now-failed) information technology company could casually remark that "you have zero privacy anyway; get over it!" (Sprenger 1999) appears on the wane. How far the pushback against surveillance has gone, how far it will go, how it will apply to health, and what results it may have, remain to be seen, however.

Recent controversies over the benefits, risks, and ultimate justifiability of surveillance have attended primarily to two areas: the use of surveillance by the state for national security and public safety, and the use of surveillance by private sector companies for marketing and other forms of economic gain. A third area has recently come under scrutiny as well: the use of big data, at least some of which has been gathered through surveillance, for political gain and influence in elections. Despite the importance of ensuring the public safety, critics see ubiquitous state surveillance as a profound threat to individual liberty. Assemblages of consumer preferences by actors in the private sector may allow advertising and information to be tailored to individuals' interests but may also target them for manipulation or price gouging. Racial, ethnic, or religious profiling may create or entrench injustice against people in disfavored groups or believed to fall into these groups. The use of social media to exercise hidden influence on elections has been seen as the newest threat to democracy. At its worst, surveillance can place individuals or groups under unjustified suspicion, locate them, shame them, cause them economic harm, marginalize them politically, and even target them for extermination.

Surveillance for public health purposes has been largely left aside in this contentious fray, at least until the appearance of COVID-19. Public health surveillance has many benefits: detecting disease outbreaks, preventing disease spread, identifying environmental toxins, improving health care, and bettering overall population health. But it shares some of the concerns that have been voiced about surveillance more generally, particularly those about threatening liberty, profiling, targeting, stigmatizing, discriminating and otherwise harming individuals or groups. Moreover, public health surveillance may not be easily isolated from surveillance that is less benign, as when information collected to improve pain management is repurposed as evidence for criminal prosecutions or immigrants' need for medical care identifies them for deportation. Indeed, public health surveillance may itself be seen as a national security measure, protecting a nation from disease or insect invasions or identifying the enemies within. When surveillance for security and surveillance for health intertwine in ways people find objectionable, however, loss of trust may be the unfortunate result. Sustaining surveillance requires maintaining this delicate balance between information needed for health and protecting people against uses they regard as objectionable or worse.

1.3   Public Health and Population Health

Surveillance may be *by* public health, *of* public health, or *for* protecting and improving public health. That is, it may be conducted by recognized public health officials, be of the health of the public, or be conducted to safeguard or better the health of the public. These are different enterprises and raise different ethical questions (e.g. Lee et al. 2010; Faden and Shebaya 2016).

In a landmark report on the future of public health in the United States, the Institute of Medicine (IOM) defined "public health" as "what we, as a society, do collectively to assure the conditions in which people can be healthy." This requires that continuing and emerging threats to the health of the public be successfully countered. According to the IOM in 1988, these threats included immediate crises, such as the AIDS epidemic; enduring problems, such as injuries and chronic illness; and impending crises foreshadowed by such developments as the toxic by-products of a modern economy (IOM 1988, p. 1). These problems are so complex and diverse, the IOM concluded, that the governmental presence in public health must reach beyond public health agencies and must be joined by the work of private organizations and individuals. At the same time, the IOM judged that the information and assessment function of public health is uniquely governmental and not to be delegated, writing that it is the duty of the public health agencies to "regularly and systematically collect, assemble, analyze, and make available information on the health of the community, including statistics on health status, community health needs, and epidemiologic and other studies of health problems" (IOM 1988, p. 7).

Views of public health that would narrow it to the activities of public health agencies argue that grants of authority to public agencies to protect and improve health must be distinguished from grants of authority to other public agencies and from activities in the private sector because of the coercive powers exercised by public health (e.g. Rothstein 2002, 2009). In advancing this view, Rothstein was especially concerned to distinguish public health from national security at a time when, after the 9/11 attacks on the World Trade towers, the United States was on high alert about potential threats from bioterrorism.

Still broader views distinguish surveillance of population health from the activities of public health agencies. Population health studies trends in the health of sets or subsets of people. Promoting population health may require addressing social determinants of health such as housing, education, structural inequality, or the protection of human rights (e.g. Goldberg 2009). Addressing the social determinants of health may require going far beyond the traditional activities of public health agencies to other governmental agencies and perhaps also to the work of many non-state actors. Non-state actors brought into play, however, may have different goals such as their own commercial ends. Moreover, these actors may be subject to different kinds of oversight than governmental agencies, raising further questions for trust in surveillance.

In the background must be the recognition that relationships between public health and the private sector are continuing to evolve. This evolution is complex and multifaceted and both local and global. Lines between clinical care and public health, personalized medicine and population health, public health and social service and activist groups, and public health and medical research are blurring and changing. The occurrence of these changes is not a new phenomenon. The relationship between public health and clinical medicine has evidenced tension throughout the ages, over explanations of disease, competencies, responsibilities, and resources. With the advent of the germ theory of disease, public health in the form of sanitation such as clean water was increasingly superseded by clinical medicine, both in the public's eye and in support from public dollars. As public health expanded case identification and contact tracing, physicians were often a source of resistance to these intrusions into their relationships with patients, as Chapter 3 describes. The Institute of Medicine in 1988 bemoaned the fragmentation of the public health system as funds were diverted away from government through tax cuts. Later writers have claimed that there have been continuing trends of public expenditures away from public health and towards health care for individuals (Tran et al. 2017). The ratio of individual health care expenditures to public health funding is particularly high in the United States; the average ratio of social expenditures to health expenditures is 2.0 in OECD countries, whereas it is .91 in the United States (Zimmerman et al. 2015). Expenditures on precision medicine efforts to tailor clinical care to individual genetic differences have raised concerns that funding will be directed away from attention to social determinants of population health, forgetting the observation of Geoffrey Rose (2001) that factors explaining sickness in individuals such as genetics may not explain variation in sickness between populations (Chowkwanyun et al. 2018; Ramaswami et al. 2018). Expenditures on health care may be expected to continue to grow as populations age, placing pressures on government budgets that may further undermine funding for public health. Indeed, funding in the U.S. Affordable Care Act that was initially intended for the Prevention and Public Health Fund was cut to pay for Medicare reimbursements to physicians and meet other expenses for ACA implementation (Keith 2018). Educational programs of public health schools are recognizing that their graduates are more likely to be working in commercial settings, non-profit settings, and other governmental agencies than in traditionally defined public health agencies (Krasna et al. 2019).

1.4 Surveillance for Health and Surveillance for Security

Surveillance may have many different goals. Specification of these goals is critical, as collection of information without the tools to make use of it is an empty enterprise (Harries et al. 2018). In a world in which gun violence is rampant and threats of bioterrorism loom, surveillance for security and surveillance for health may understandably seem interconnected. Pressures may intensify to share data about individual mental health for public safety and about diagnoses of unexpected health events for national security. Information about health may be put to other purposes as well, such as finding and deporting undocumented immigrants. Responses to these pressures may serve essential health-protective goals, such as detecting COVID-19 outbreaks, release of anthrax spores, or sarin gas. But they may also reflect fear, demonization of the other, and political repression. When public health fails in either direction— failing to keep people safe or over-reaching in the pursuit of aggrandizement—it may no longer be trusted by the public or important sectors of it. The justifiable—and likely realistic—result will be the loss of trust in surveillance and its eventual instability. Hence the title of this volume, "sustaining surveillance": our goal is to explore the ethical problems that can undermine surveillance and the promises that can enable it to continue reliably and responsibly in the context of warranted trust.

In U.S. law, the power of the state to protect the public health has been cemented against individual rights. Individual rights are often challenged as the apparent gravity of a public health threat grows. In the 1873 *Slaughterhouse Cases* (1873), the Supreme Court limited the privileges and immunities clause of the newly-adopted Fourteenth Amendment to rights in the U.S. constitution, permitting the state to establish a business monopoly for a slaughterhouse that agreed to comply with what the state claimed were health and safety standards. The Supreme Court has upheld the authority of the state to compel vaccination (*Jacobson* 1905) and sterilization of those believed to be intellectually disabled, however erroneously (*Buck* 1927). In the latter case, which is heavily criticized but has never been overruled, Justice Holmes opined, "the public welfare may call upon the best citizens for their lives" (*Buck* 1927, 207). The constitutional protection against arbitrary searches and seizures has analogized protection of public health to protection of national security (Fairchild et al. 2007, 16). In a quickly overruled decision upholding the warrantless search of a home for rat infestation by a health inspector, the Supreme Court said that the inviolability of the home must yield when the community seeks to maintain minimal standards of health and wellbeing (*Frank* 1959). Fairchild and colleagues also describe the interrelationship between concerns about biological warfare and the development of surveillance systems during the Cold War (2008, 17). Since the 9/11 attacks on the World Trade towers, mandated by the Public Health Security and Bioterrorism Preparedness and Response Act of 2002, the U.S. has continued to develop systems for syndromic surveillance to sense both potential bioterrorism and outbreaks of disease (e.g. Gould et al. 2017).

The analogy between health and security also has been invoked on the side of protecting liberty. The decision to allow the rat-catcher to enter the home without a warrant was met by a vehement dissent from four justices—Chief Justice Warren and Justices Black, Brennan, and Douglas—who argued that allowing arbitrary searches and seizures by the health department would also allow the government to conduct arbitrary searches for subversives (*Frank* 1959). The decision was overruled a mere eight years later with the ascendancy of the liberal majority of the Warren Court, which determined that health department searches to abate a public nuisance required a warrant (*Camara* 1967). This decision did not abate the ultimate authority of the state to protect health and security, however; it only subjected that authority to the guarantees of due process. In the only Supreme Court decision regarding requirements to report individual health data to public health, the Court upheld the reporting requirements so long as the data were adequately protected from disclosure. This decision is highly relevant to surveillance activities today, as it involved the state's creation of a data base of controlled substance prescriptions to avoid drug misuse (*Whalen* 1977).

Debates about access to firearms in the United States are a particularly telling example of how surveillance for public health purposes and surveillance for public safety may intertwine and raise troubling issues of profiling, discrimination, and individual liberty. Gun violence is both a public health issue and a public safety issue. In the wake of mass shootings in the U.S., proposals have been pressed to identify people with diagnosed mental illness and restrict their ability to purchase guns. Advocates for people with mental illness strongly oppose these proposals, on grounds that they violate the rights of people with mental illness, that people with mental illness are very unlikely to present risks of violence, and that in fact people with mental illness are far more likely to be victims than perpetrators of violence. Advocates also point out that access to mental health services is woefully lacking for many in the U.S. Increasing access to these services would be both fairer and more effective preventive measures, mental health advocates claim. Moreover, people who do pose risks of violence may be difficult to identify, whether or not they are also mentally ill. Into this mix, proposals have surfaced to the effect that data analytic techniques might be able to predict people who are at risk of committing violent acts and whose immediate access to guns should therefore be limited. Similar proposals have surfaced about possibilities of identifying people who are at imminent risk of suicide. These proposals of course raise challenging questions about reliability and the risks of both false positives and false negatives. But with increasing accuracy they also bring into sharp focus the interplay between public health and security and the ethical challenges of both furthering the overall public good and respecting the individuals who make up that public.

1.5   Framing the Ethics of Public Health Surveillance

This book stands at the intersection of major theoretical developments in the ethics of public health, the ethics of data use, and philosophy of public policy. Public health ethics is no longer primarily utilitarian. Fair information practice principles are undergoing reevaluation in light of the exploding world of big data and learning health care systems. The philosophy of public policy is developing accounts of the relationship between ideals and the imperfections of a world of structural injustice and individual failures to act justly. A brief survey of some of these developments is helpful for understanding the discussion to follow of more specific surveillance practices.

Until about the last quarter-century, public health ethics was seen primarily as utilitarian: what means would be most effective in enhancing the overall health of the public? Human rights—whether political, economic, or social—were placed at best orthogonally to the goal of overall public health, at times furthering it but at times standing in the way. Individual rights—to informed consent, privacy and confidentiality, or to decline or even receive treatment—could present roadblocks to the overall improvement of public health. To be sure, rights protection might be of instrumental value, if assurance that their rights were protected would encourage people with dangerous infections to seek care that could prevent or retard contagion. But rights might also stand in the way of requiring people to share information or to undergo needed treatment. Spurred in particular by the efforts of the late Jonathan Mann against HIV/AIDS during the 1990s, public health ethics has changed significantly. For example, Nancy Kass, in one of the most influential frameworks for the analysis of ethical issues in public health, balances achieving public health goals of furthering health against privacy, fairness, and liberty and self-determination (Kass 2001).

As public health ethics has been developing accounts of individual rights and fairness in addition to pursuing goals of protecting and promoting health, the critical need for effective surveillance to identify and hopefully prevent pandemic disease in a global world has come to the center of recent international health policy. In 2005, the World Health Organization (WHO) adopted new International Health Regulations. In effect since 2007, the Regulations give the WHO far broader authority to declare public health emergencies of international concern. They also impose on states parties the obligation to notify WHO of events that might constitute these emergencies and to maintain surveillance capacities adequate "to detect, assess, notify, and report events in accordance with these Regulations" (WHO 2005, Art. 5(1)). States parties are also to make efforts to collaborate with each other to the extent possible in detecting and responding to these events, providing technical and logistical support, and mobilizing financial resources to develop their surveillance obligations. These Regulations have been put to the test with at best mixed success in several recent events, most notably the 2009 H1N1 influenza pandemic and the outbreak of Ebola in West Africa in 2014–2016. If the United States had continued the strategy of withdrawing from the WHO according to the notice given by President Trump, the impact of the Regulations could have been further strained.

In late 2017, the WHO issued guidelines on ethical issues in public health surveillance (WHO 2017). Characterizing surveillance as the "radar" of public health, the Guidelines (at 14) define surveillance broadly as "the continuous, systematic collection, analysis and interpretation of health-related data needed for the planning, implementation, and evaluation of public health practice." Such surveillance may include not only disease and injury, but also important public events and environmental conditions that may affect health, as well as vital statistics that may reveal health trends. The Guidelines specifically contrast research (including epidemiological research), where the rights paradigm of individual informed consent holds sway, with public health surveillance where values such as the public good and social solidarity take precedence. The Guidelines rely on a backbone of four ethical considerations—the common good, equity, respect for persons, and good governance—to develop a list of seventeen more specific guidelines. These guidelines include topics such as the obligation to develop effective surveillance capabilities, limits on the purposes for which data may be used, transparency, global obligations of support, respect for community values, harm mitigation and risk disclosure, data integrity and security, obligations on the part of individuals to contribute information, effective communication of results, and data sharing.

The Guidelines are justifiably characterized by those involved in their creation as the first systematic effort to assess surveillance from the perspective of public health ethics (Fairchild et al. 2017). But they are only a beginning, useful but cautious. They are limited in focus to surveillance by public health, but surveillance for public health can extend far beyond what public health agencies do. Their primary goal is ethical development of the infrastructure needed to detect emergencies—certainly a critical goal, but not the only one. Although extensive, this list of guidelines is skeletal at best and contains little analysis of how ethical considerations might be balanced when guidelines point in different directions. Moreover, as the authors recognize, surveillance capacities and methods are rapidly changing in this era of big data. In what follows, we will draw on and refer to this very useful document where relevant.

Information ethics have also been developing rapidly since the dawn of the computer age. In 1973, stimulated by concerns about automated data systems, the U.S. Department of Health, Education, and Welfare published a report titled *Records, Computers, and the Rights of Citizens* (DHEW 1973). The report developed a set of fair information practice principles and urged that they become the backbone of an all-encompassing U.S. privacy law. These principles included transparency about data collection, protection for the individual's ability to find out what information was being collected and to correct errors, and the requirement that data not be collected for one purpose and used for another without the individual's consent. Although the upshot of the HEW report was only a US federal Privacy Act applying to agencies of the federal government, FIPs, as they have become known, have been highly influential on the development of data privacy policies ever since, from the Data Protection Regulation of the European Union to the Health Insurance Portability and Accountability Act security and privacy rules for protecting health information in the United States (Gellman 2019).

In the years after their initial formulation, FIPs have been challenged as both over- and under-protective. On the side of what is arguably over-protective in light of public health data needs, FIPs rely on a model of individual notice and consent for data collection and use. This model is highly constricting for public health uses of data, both because of the time and resources involved and because of the likelihood that some individuals will value privacy so highly that they will refuse to consent to any uses of data beyond their own health care. Both US and EU regulatory structures permit use of individual information for public health without the notice and consent of FIPs, to the extent required by law. This exception for public health might be judged as necessary for protection or as a giant loophole allowing the state to run roughshod over individual rights. Or, the conclusion might be drawn that it is too extreme either to give public health free rein over the use of information if legislatively permitted or mandated, or to require individual informed consent for data collection or use.

An intermediate solution has been to distinguish information about individuals that is thought to be identifiable and information that is anonymous or deidentified.

Many regulatory structures about the collection or use of information, including regulations governing research with human subjects, draw this line. With the advent of complex data sets and increasingly sophisticated statistical tools, however, the line between identification and deidentification has come under question. When data sets are combined, the enriched information they contain may allow significant percentages of individuals to be reidentified. Some argue in addition that uses of fully anonymized data may be problematic, if it allows groups to be stigmatized or inferences to be drawn about individuals in virtue of their membership in groups. Another solution is to rely more heavily on the FIPs of transparency about data collection and use and about purpose limitation to reassure individuals about the appropriateness of data use.

Finally, debates in the philosophy of public policy have come to grips with our world of structural injustice and individual failures to behave justly. Spurred by John Rawls's account of an ideal of liberal justice (1971), which famously set aside what justice might require in less than favorable circumstances, theorists have developed "non-ideal" or "partial compliance" approaches. These take many forms but ask the general question whether the requirements of justice are different in circumstances of injustice than in circumstances where justice prevails.

Much recent discussion in political philosophy has attended to claims that Rawlsian ideal theory neither can nor should be applied to the real world of serious injustice. In a very helpful conceptual map of this terrain, Laura Valentini (2012) distinguishes three areas of theorizing about departures from the ideal: obligations of some when significant numbers of others are failing to fulfill their obligations, the extent to which feasibility of realization should constrain normative political philosophy, and requirements of justice in transitioning from the non-ideal to the ideal.

Each of these areas of non-ideal theory raises central questions for the ethics of public health surveillance. What are the obligations of some states when other states are refusing to comply with ethical requirements, either to provide data or to support their efforts to do so? If some states suppress urgently needed information about outbreaks of contagion, may others respond by refusing to help with needed medical equipment or vaccines, or even by withdrawing from international cooperative organizations? What are the obligations of individuals to allow information about themselves to be shared when others are hiding information that might be critical to their health? How do practical constraints such as costs, willingness to comply, or the availability of information, affect what surveillance should ethically do? Should protection of individual rights and liberties be weighed differently in the effort to ameliorate structural injustices that are damaging health than they would be weighed in more favorable circumstances? Some believe that ideal justice is at least relevant to answering questions about what to do in an imperfectly just world. Others believe that theorizing about justice in non-ideal contexts is simply different from theorizing in ideal contexts and should not be guided by images of ideal justice (e.g. Wolff 2017; Sen 2009).

This volume is rooted firmly in the territories of non-ideal theory. As such, we do not assume that a complete and final framework for the ethics of surveillance and public health should be our goal. Rather, as partial compliance theorists we believe it is important to proceed on two fronts: identifying particular ethical pitfalls to be avoided and understanding promises it is realistic to pursue. These strategies of non-ideal theory structure our book.

1.6   Core Ethical Considerations for Surveillance

Data collection and use for public health must always be understood as a balance between some individuals or groups and others. Although public health as practiced by public agencies tries to be health for all, and to be so equitably, what it can do is limited. Resources, including time and talent, are in short supply. Private actors may supplement public health surveillance but may have different goals, for example deciding where to locate businesses in a community, what products to develop and how to price them, or how to manage a labor force. Discussion of the ethics of these uses and users of information for public health must include benefits, risks, and equities. Uses of data that are unexpected or disconcerting may destabilize surveillance. Destabilization is especially likely if surveillance by private actors for commercial gain is blended or confused with surveillance by public agencies for the overall good. Destabilization may also occur with uses of data that offend, that generate resentment, or that demean. Uses of data that cause direct harm—or the fear of direct harm or even the anecdote or urban legend of direct harm—may also threaten the continued integrity of surveillance. When private actors enter the mix, there may be much less control about transparency, about data that are collected and stored, and about how data are analyzed and used.

Here, then, are some basic considerations to apply to any forms of surveillance for public health. These considerations begin with some of the Fair Information Practice principles (FIPs), but go beyond them in ways that are explained as the argument of this volume develops.

Transparency. Surveillance should always be publicly known. What information is collected, how it is processed and maintained, and who is responsible for the information should be matters of public record. This does not mean that individual pieces of information should be public; that information might compromise the individuals it concerns. Rather, it only means that there should be no secret surveillance operations.

Purpose specification. Information collected by and for public health should be used for health. This can include research about health improvement; there is no clear line between public health activities and public health research. Information for public health should not be diverted to other purposes, such as national security or commercial advantage, without further authorization. If appropriate, such further authorization may mean individual or community consent.

Harm minimization. Any risks to individuals and groups from data collection and use should be identified and minimized to the extent possible. If harms are unavoidable, compensation should be considered, including in the form of access to health care.

"Giving back." In what follows, we will be arguing that at least some information, even information about individuals, should be shared with public health. Health is a collective, not merely an individual good—as societies have learned to their regret with COVID-19. Information for public health is a social responsibility. This does not mean that all information about individuals is within the purview of public health, but it does mean that individual consent is not always needed for public health access to information. It does mean, however, that this access should only be conducted within appropriate purpose specification. Moreover, giving back goes both ways. To the extent possible, individuals should be made aware of what is being achieved when information is shared with public health. Such sharing of the benefits is critical to enlisting ongoing support for public health efforts.

Equity. Burdens and benefits of public health should be equitable. Public health has responsibilities to address health inequities and social determinants of health to the extent that they contribute to health inequities. Public health that works for some but not for others will not be sustainable in the long run.

These principles are stated quite abstractly and must be applied in nuanced ways. We will explain and defend them over the course of this volume. However, unless attention is paid in some important ways to each of them, surveillance may not be met with warranted trust and ultimately may not be sustained.

1.7 Plan of the Volume

We start in Part I with traditional forms of surveillance: counting population numbers, finding cases of deadly disease and tracing their contacts, and unearthing toxins in the environment. Each of these forms of surveillance has been and remains an important part of the armamentarium of public health. Yet each also reveals problems that must be avoided if surveillance is to be sustainable. They have been joined by forms of technologically enhanced data collection and use that raise further problems of privacy and injustice.

Counting population numbers allows governments to assess trends in the health of the population and to identify the ravages of disease. Yet counting numbers can also lead to stigmatization of populations, resistance, and suspicion of what the numbers are thought to yield. One lesson to draw from historical mistakes such as the identification of native Hawai'ians with leprosy, of Chinese with plague, or of Haitians with HIV is the need for trust in and trustworthiness of science and how it is communicated. Chapter 2 explores counting population numbers and some pitfalls of ethics in the science of surveillance.

Finding cases of deadly disease and tracing their contacts can stop disease spread. But it can also impose immediate risks on the individuals thus identified. It can threaten their physical security, personal relationships, economic security, employment, and even health. It can violate basic rights and—as Chapter 3 describes—result in calls for protection of rights that may backfire on public health. Contact tracing that is enhanced by locational tracking or other uses of smart phones may exacerbate these tensions between halting disease and protecting individuals. Case finding and contact tracing also must be coupled with the recognition that potential spreaders may require support for their needs and their health.

Before the role of microorganisms in causing disease was understood, public health sought to address environmental uncleanliness: odors, sewage, and the "miasma" of bad air. With the germ theory of disease came the focus on individuals as transmitters of disease, to the detriment of attention to the environment. Yet, as Chapter 4 details, decisions about what to surveille can be as important to ethical surveillance as decisions about how to surveille. Failures to attend to inequities in surveillance can undermine willingness to participate. When some believe that data about themselves is being used solely to the benefit of others, resentment is an understandable result.

Enhanced computing power and the advent of the internet have brought new forms of data, new methods of collecting that data, and new forms of storing and analyzing that data. It has also brought powerful new private actors into surveillance efforts, both for commercial purposes and for public health. Chapter 5 considers the issues of privacy and justice that these new methods raise.

Chapter 6 takes up what has been called the "new" public health. In the world today, non-contagious diseases take a far greater toll on health than contagious diseases. Public health has understandably reached out to surveille and address these threats to health. Obesity, lack of exercise, poor diet, alcohol and substance abuse, all cause ill health. Yet intervention with individuals' choices about how to live their lives has been portrayed as the "nanny" state and criticized as unjustified paternalism. On the other hand, many of these so-called diseases of despair are rooted in social factors beyond the individual. Here, we argue, public health surveillance must take care not to over- promise and over-reach if its efforts are to be regarded as warranted. Instead, it must proceed with respect for persons, acknowledging that people may have different values including about the importance of health.

In reaction to paternalism and violations of individual rights, paradigms of informed consent from bioethics and research ethics have been brought into discussions of the ethics of surveillance. Some insist that consent to data use is both necessary and sufficient for permissible surveillance, along the lines of strict views of FIPs. Models for opting in, or opting out, of data use have been developed, as have models of group or community consent. In Chapter 7, we consider the ethical imperatives behind these models, of transparency, participation, and respect for individuals as determining their own conceptions of their good. We suggest how these imperatives may be recognized without full requirements for consent and the barriers these might pose for public health.

In the final chapter, we bring our earlier discussions together by seeing what can be learned from the critical failures of surveillance for COVID-19. In many ways, societies had lost sight of the importance of surveillance. There were continuing failures to address the core questions we have identified that encourage people to support, participate in, and believe in the information provided by surveillance. Only by continuing to consider and provide answers to these questions can trust in data use be warranted and surveillance ultimately sustained. Establishing trust in data use is only part of the picture, however. Ultimately, people must also see the benefits of data use for their own lives if support for surveillance is to be sustained.

References

Barry, John M. 2004. *The great influenza: The story of the deadliest pandemic in history*. London: Penguin Books.

Bridgman, Aengus, Eric Merkley, Peter John Loewen, Taylor Owen, Derek Ruths, Lisa Teichmann, and Oleg Zhilin. 2020. The Causes and Consequences of COVID-19 Misperceptions: Understanding the Role of News and Social Media. Harvard Kennedy School Mis/information Review (June 18) [online] https://misinforeview.hks.harvard.edu/article/the-causes-and-consequences-of-covid-19-misperceptions-understanding-the-role-of-news-and-social-media/. Accessed 10 July 2020.

Buck v. Bell, 274 U.S. 200 (1927).

Camara v. Municipal Court of the City and County of San Francisco, 387 U.S. 523 (1967).

Carpenter v. United States, 585 U.S. ___, 138 S. Ct. 2206, 201 L. Ed. 2d 507 (2018).

Chowkwanyun, Merlin, Ronald Bayer, and Sandro Galea. 2018. "Precision" Public Health—Between Novelty and Hype. *New England Journal of Medicine* 379 (15): 1398–1400.

Department of Health, Education, and Welfare. 1973. Records, Computers and the Rights of Citizens: Report of the Secretary's Advisory Committee on Automated Personal Data Systems July 1973. Electronic Privacy Information Center. https://epic.org/privacy/hew1973report/default.html. Accessed 11 July 2020.

EUR-Lex. 2016. REGULATION (EU) 2016/697 OF THE EUROPEAN PARLIAMENT AND OF THE COUNCIL of 27 April 2016 on the Protection of Natural Persons with Regard to the Processing of Personal Data and on the Free Movement of Such and Repealing Directive 95/46/EC (General Data Protection Regulation). https://eur-lex.europa.eu/legal-content/EN/TXT/?uri=celex:32016R0679. Accessed 11 July 2020.

Faden, Ruth R., and Sirine Shebaya. 2016. Public Health Ethics. In *The Stanford Encyclopedia of Philosophy* (Winter 2016 Edition), ed. Edward N. Zalta. https://plato.stanford.edu/archives/win2016/entries/publichealth-ethics/. Accessed 1 Aug 2020.

Fairchild, Amy L., Daniel Wolfe, James Keith Colgrove, and Ronald Bayer. 2007. *Searching Eyes: Privacy, the State, and Disease Surveillance in America*. Berkeley: University of California Press.

Fairchild, Amy L., Ronald Bayer, and James Colgrove. 2008. Privacy, Democracy and the Politics of Disease Surveillance. *Public Health Ethics* 1 (1): 30–38.

Fairchild, Amy L., Angus Dawson, Ronald Bayer, and Michael J. Selgelid. 2017. The World Health Organization, Public Health Ethics, and Surveillance: Essential Architecture for Social Well-Being. *American Journal of Public Health* 107 (10): 1596–1598.

Frank v. Maryland, 359 U.S. 360 (1959).

Gellman, Robert. 2019. Fair Information Practices: A Basic History. https://bobgellman.com/rg-docs/rg-FIPshistory.pdf. Accessed 10 July 2020.

Gmeinder, Michael, David Morgan, and Michael Mueller. 2017. How Much Do OECD Countries Spend on Prevention? OECD Health Working Paper No. 101. https://www.oecd-ilibrary.org/docserver/f19e803c-en.pdf?expires=1562104557&id=id&accname=guest&checksum=53EB160DDC6B0CDD6BC1F876D167175A. Accessed 2 July 2019.

Goldberg, Daniel S. 2009. In Support of a Broad Model of Public Health: Disparities, Social Epidemiology, and Public Health Causation. *Public Health Ethics* 2 (1): 70–83.

Gould, Deborah W., David Walker, and Paula W. Yoon. 2017. The Evolution of BioSense: Lessons Learned and Future Directions. *Public Health Reports* 132 (aSuppl): 7S–11S.

Harries, A.D., M. Khogali, A.M.V. Kumar, S. Satyanarayana, K.C. Takarinda, A. Karpati, P. Olliaro, and R. Zachariah. 2018. Building the Capacity of Public Health Programmes to Become Data Rich, Information Rich and Action Rich. *Public Health Action* 8 (2): 34–36.

Honigsbaum, Mark. 2019. *The pandemic century, one hundred years of panic, hysteria and hubris.* New York: W. W Norton and Company.

Institute of Medicine. 1988. *The Future of Public Health.* Washington, DC: National Academies Press.

Jacobson v. Massachusetts, 197 U.S. 11 (1905).

Jamieson, Kathleen Hall, and Dolores Albarracin. 2020. The Relation Between Media Consumption and Misinformation at the Outset of the SARS-CoV-2 Pandemic in the US. The Harvard Kennedy School Misinformation Review (April) [online]. https://misinforeview.hks.harvard.edu/wp-content/uploads/2020/04/April19_FORMATTED_COVID-19-Survey.pdf. Accessed 10 July 2020.

Kass, Nancy E. 2001. An Ethics Framework for Public Health. *American Journal of Public Health* 91 (11): 1776–1782.

Keith, Katie. 2018. New Budget Bill Eliminates IPAB, Cuts Prevention Fund, and Delays DSH Payment Cuts. *Health Affairs Blog* (Feb. 9). https://www-healthaffairs-org.ezproxy.lib.utah.edu/do/10.1377/hblog20180209.194373/full/. Accessed 4 Aug 2020.

Krasna, Heather, Julie Kornfeld, Linda Cushman, Shuyue Ni, Pantelis Antoniou, and Dana March. 2019. The New Public Health Workforce: Employment Outcomes of Public Health Graduate Students. *Journal of Public Health Management and Practice.* epub ahead of print. https://doi.org/10.1097/PHH.0000000000000976.

Lee, Lisa M., Steven M. Teutsch, Stephen B. Thacker, and Michael E. St. Louis, eds. 2010. *Principles and practice of public health surveillance* (3rd ed.). New York: Oxford University Press.

OECD. 2019. Health Spending (Indicator). https://doi.org/10.1787/8643de7e-en. Accessed 10 July 2020.

Ramaswami, Ramya, Ronald Bayer, and Sandro Galea. 2018. Precision Medicine from a Public Health Perspective. *Annual Review of Public Health* 39: 153–168.

Rawls, John. 1971. *A Theory of Justice.* Cambridge, MA: Harvard University Press.

Rose, Geoffrey. 2001. Sick Individuals and Sick Populations. *International Journal of Epidemiology* 30: 427–432.

Rothstein, Mark A. 2002. Rethinking the Meaning of Public Health. *Journal of Law, Medicine and Ethics* 30: 144–149.

———. 2009. The Limits of Public Health: A Response. *Public Health Ethics* 2 (1): 84–88.

Sen, Amartya. 2009. *The Idea of Justice.* Cambridge, MA: Harvard University Press.

Simonov, Andrey, Szymon K. Sacher, Jean-Pierre H. Dubé, and Shirsho Biswas. 2020. The Persuasive Effect of Fox News: Non-Compliance with Social Distancing During the Covid-19 Pandemic. NBER Working Paper No. 27237 (June). National Bureau of Economic Research. https://www.nber.org/papers/w27237. Accessed 10 July 2020.

The Slaughterhouse Cases, 83 U.S. 36 (1873).

Sprenger, Polly. 1999. Sun on Privacy: 'Get Over It'. *WIRED* [online] (Jan. 26). https://www.wired.com/1999/01/sun-on-privacy-get-over-it/. Accessed 10 July 2020.

Suthar, Amitabh B., Lisa G. Allen, Sara Cifuentes, Christopher Dye, and Jason M. Nagata. 2018. Lessons Learnt from Implementation of the International Health Regulations: A Systematic Review. *Bulletin of the World Health Organization* 96: 110–121.

Tran, Linda Diem, Frederick J. Zimmerman, and Jonathan E. Fielding. 2017. Public Health and the Economy Could Be Served by Reallocating Medical Expenditures to Social Programs. *SSM—Population Health* 3: 185–191.

Valentini, Laura. 2012. Ideal vs. Non-ideal Theory: A Conceptual Map. *Philosophy Compass* 7 (9): 654–664.

Whalen v. Roe, 429 U.S. 589 (1977).

Wolff, Jonathan. 2017. Forms of Differential Social Inclusion. *Social Philosophy and Policy* 34 (1): 164–185.

World Health Organization (WHO). 2005. International Health Regulations. http://apps.who.int/iris/bitstream/handle/10665/246107/9789241580496-eng.pdf;jsessionid=623AB24B3BB0E8CF92F14D7956DBDF12?sequence=1. Accessed 11 July 2020.

———. 2017. WHO Guidelines on Ethical Issues in Public Health Surveillance. http://apps.who.int/iris/bitstream/handle/10665/255721/9789241512657-eng.pdf?sequence=1.   Accessed   11 July 2020.

———. 2020. Timeline of WHO's Response to COVID-19 (last updated 30 June 2020). https://www.who.int/news-room/detail/29-06-2020-covidtimeline. Accessed 10 July 2020.

Zimmerman, Emily B., Steven H. Woolf, and Amber Haley. 2015. *Understanding the Relationship Between Education and Health*. Washington, DC: Agency for Healthcare Research and Quality. https://www.ahrq.gov/professionals/education/curriculum-tools/population-health/zimmerman.html. Accessed 31 July 2020.

Chapter 2
Counting Numbers

2.1 Background

Surveillance began with counting the numbers of people in the population. Numbers were used to assess the overall strength of the population and to identify the march of dangerous contagion. Taking the census was how kings or emperors—or their enemies—could determine whether their subjects were strengthening or weakening in numbers. Such census-taking was an early important source of vital statistics and thus to some extent of the identification of public health trends. As the statistician Carlos Grajales recounts in a history prepared for the Royal Society of Statistics, counting numbers was how the Babylonians determined grain storage needs to feed the population about six thousand years ago and how the Egyptians calculated the labor supply required to build the pyramids. The book of numbers in the *Bible* records the counting of the Israelites and their ability to bear arms during the flight from Egypt. The Romans conducted a census every five years, overseen by a high official known as the censor. The Han dynasty census, conducted in AD 2, is the earliest known full census record (Grajalez 2013). The Domesday Book, commissioned by William the Conqueror in 1085, recorded lands and landholding in the England of the day. Periodic inventories of land and inhabitants were taken thereafter.

Vital statistics include births, deaths, and where they occurred. They may also include causes of death to the extent known, marriages, and other information. These data may serve many social needs beyond health, including determination of citizenship, establishment of legal parentage, and processing of insurance or inheritance claims (Dunn 1936). Vital statistics may be kept nationally or locally and are often maintained by religious organizations as well as by the state.

Recording epidemic outbreaks began about as early as collection of census data. Historians date the first recorded epidemic to a plague in Egypt in 3180 B.C. (Choi 2012). Subsequent outbreaks of plague were documented during the time of the Roman emperor Justinian and during the fourteenth century in the fruitless hope of halting disease spread. Until the last several hundred years, vital statistics, complemented by news of sickness, were the primary means for gathering knowledge

© Springer Nature Switzerland AG 2021
J. G. Francis, L. P. Francis, *Sustaining Surveillance: The Importance of Information for Public Health*, Public Health Ethics Analysis 6,
https://doi.org/10.1007/978-3-030-63928-0_2

about locations and progress of pestilences such as plague or cholera. These records were aimed at a foremost goal of public health surveillance: detecting outbreaks of deadly disease. However, they often were not sufficiently timely to allow societies to take action against the disease as it spread; instead, they served primarily as historical records of the disease's impact.

Outbreak detection presents a compelling case for timely and comprehensive information to avoid, mitigate, or—most hopefully—extinguish the spread of sickness. Counting cases of illness and death within a population can reveal the severity and distribution of an outbreak, as well as possibilities for exposure as the outbreak spreads to epidemic or pandemic levels. With COVID-19, daily numbers of reported new cases, deaths, and reportedly "recovered" patients who have survived at least three weeks from diagnosis, serve to remind political leaders and the public of the pandemic's toll. Estimates of "excess" deaths—numbers of deaths beyond what would normally have been expected within a population—are also used to assess the pandemic's hidden impact. Vital statistics also reveal the pandemic's disparate impact on the elderly, people of color, people with disabilities, and the poor.

But even simple counting of population numbers, vital statistics, or reports of disease has been controversial. Forceful opposition to systematic counting of the population has persisted throughout history. Some judged that counting the population was sacrilegious because of fears that it might incur the wrath of God (NISRA 2019). Others were concerned that the results would reveal the country's weaknesses to its enemies or that gathering the information was a threat to individual liberties. Still others resist collecting certain kinds of demographic information about the population; for example, France does not collect racial and ethnic categories in its census (Léonard 2015).

These controversies raise some of the most fundamental issues about surveillance. Information is power, and the most rudimentary surveillance can be used both for good and for harm. Knowledge of population numbers has enhanced the power of monarchs to levy oppressive taxes or conscript soldiers. It has been regarded as sacrilegious in some religious traditions. It has been deployed to reveal migration patterns or to find immigrants themselves, as well as in the U.S. to allocate political representation in ways that have been unfair at times or clearly discriminatory. It has been thought to reveal generalizations about population subgroups that may be stigmatizing or degrading. Political leaders may also wish to suppress information about numbers or present numbers in ways that make them appear more favorable than they actually are. Rulers may hope to moderate alarm among the population or to enhance their political positions. President Trump's decision in July 2020 to have hospitals report COVID-19 data to the Department of Health and Human Services rather than to the Centers for Disease Control and Prevention was criticized by those who feared exactly this kind of data manipulation for political gain (Stolberg 2020).

This chapter sets ethical questions about these basic surveillance methods in historical and epistemological context. It gives examples of uses of data about population numbers, vital statistics, or outbreaks that have been clearly beneficial, as well as examples that have bordered on the genocidal. Counting numbers, as a rudimentary epidemiological method, also presents the opportunity to consider some of the problems raised by epidemiology as a science and the ethical implications of how these problems are answered. How can the science of epidemiology—or science more generally—go wrong in ways that might undermine ethical justifications for surveillance?

2.2 Plagues and Pandemics: From the Black Death to COVID-19

The spread of bubonic plague presents perhaps the most sustained examples of counting cases of disease over the centuries. Recent outbreaks of contagion have been counted, too, from Ebola to COVID-19. Daily logs of new infections and deaths were published as the COVID-19 pandemic spread.

Common themes are reflected in these numbers and how they are publicized. The roles of fear, misunderstanding, and mistrust are apparent. So is the perception of contagious disease as attacking from without and the use of military rhetoric in describing responses to its spread. Some communications have stigmatized disease victims. Others have emphasized the interactions between health disparities and burdens of infection. Finally, people seem to care about having information that is timely and complete. All of these themes can be found in the history of the plague.

2.2.1 The Plague

Plague, caused by the bacterium *Yersinia pestis*, stimulated collection of vital statistics. Plague is deadly, with fatality rates of 30% to 100% without treatment. Plague killed more than fifty million people in Europe alone during the fourteenth century, over half of the population at the time. Plague is also ugly: its symptoms include weakness, seizures, diarrhea and vomiting, bleeding, swollen lymph nodes, and the blackened skin that gave it the label "Black Plague." For most of history, all that people could do for protection was to become aware of where disease had broken out and make efforts to avoid it; today, early antibiotic therapy can successfully treat most cases of plague. How plague spread—through bites of infected fleas or lice—was not known until the last century, either. Even today, how plague spreads remains an ongoing subject of study; until recently, rats were thought to be the vehicles transporting fleas, especially aboard ships, but modeling now suggests human transport was to blame for carrying plague-bearing fleas across the globe (Dean et al. 2018).

The plague also illustrates the range of human reactions in the face of pandemic spread. Immediate threats of horrifying and deadly diseases are commonly met by panic and fear. Plague was recognizable, disgusting, and deadly. All too often, fearful reactions to such diseases trace racial or ethnic lines, judgments of moral opprobrium, or both. Reactions may condemn or destroy cultural or religious practices. Those supposed to be victims of disease may be seen as deadly sources of contagion and quarantined, banished, or exterminated.

As the plague swept apparently inexorably across Europe during the early Renaissance, strategies to prevent disease spread developed, such as quarantine and the cordon sanitaire. Quarantine—the practice of keeping ships offshore for forty days until all disease was supposed to have died out—was instituted in the region of Venice in 1377 to protect against plague (Tognotti 2013; Gensini et al. 2004). The Venetian Republic appointed three guardians to detect and exclude ships carrying the disease (Declich and Carter 1994). The first English quarantine regulations were adopted in 1663 in London and the initial French regulations in 1683 in the port city of Marseille—both also to stop plague from entering the city. A further strategy to halt disease spread was drawing geographical lines that could not be crossed—the so-called cordon sanitaire. The heroic village of Eyam in England self-isolated, perhaps creating even greater risk for its residents by transforming their disease to its deadlier pulmonary form (Massad et al. 2004).

Plague arrived on the west coast of the United States towards the end of the nineteenth century, apparently from China via Hawai'i during a pandemic that originated in southern China and spread widely in Asia and Europe. The disease was first found in Hawai'i among Chinese residents. At the time, although the bacillus causing the disease could be identified, its mode of transmission was unknown. The assumption was common that white European ancestry conferred immunity (Randall 2019, p. 6). Honolulu's Chinatown was quarantined out of fear of the pestilence. When the home of one of the victims was burned to eradicate the infection, a shift in the winds flared the fire out of control and Honolulu's entire Chinatown was reduced to ashes.

When a case of plague was identified in San Francisco's Chinatown a few months later, it appeared in a city where anti-Chinese sentiment was fierce. Prejudice and fear combined to impose immediate quarantine on all of Chinatown. But quarantine competed with corruption and concerns about its economic impact on the city. Thereafter, efforts to address the plague ricocheted between quarantine and release amidst disbelief in bacteriological confirmation of the disease. (Suspicions of science are not merely a phenomenon of the present day.) Evidence was also clear that bodies were being hidden for fear of discovery (Randall 2019, p. 58). Efforts were imposed to prevent movement outside of the state by people of Chinese ancestry who had not received an experimental vaccine. These impositions were enjoined by the courts as violating constitutional rights to equal protection (*Wong Wai* 1900; *Jew Ho* 1900) because no evidence had been provided for imposing the restrictions only on Asian residents of San Francisco (McClain 1986). From the outbreak in San Francisco, plague spread across the bay and to Los Angeles and beyond; the disease is now endemic throughout the western United States. Ignorance about disease transmission, prejudice, and economic protectionism all contributed to this disease spread and the failure to prevent its becoming endemic in the U.S.

Today, plague is found on all continents, although the three most affected countries are the Democratic Republic of Congo, Madagascar, and Peru. According to the World Health Organization (WHO 2017a), surveillance is essential to identify and manage plague outbreaks wherever they might occur. Likewise, surveillance is critical to detecting outbreaks of polio, cholera, Ebola, avian influenza, and the myriad other infectious diseases known in the world today—along with emerging infections as yet unknown. Counting numbers of people who are sick and dying still matters today to this enterprise, although far more sophisticated surveillance techniques are also in use.

2.2.2 Ebola

Ebola is a relatively new zoonosis, a disease initially transmitted from non-human animals to humans. Because people with Ebola bleed copiously, caregivers for them are at high risk of infection without effective precautions. Burial practices that involve bathing or dressing infected corpses are also highly dangerous as they may involve contact with infected fluids. Only recently have vaccination or treatment been available for this once highly deadly disease (Farmer 2020; Maxmen 2010). Timely information is thus critical to prevent Ebola spread, yet the history of identifying Ebola outbreaks is a history of surveillance challenges and failures.

The 2014–2016 outbreak in West Africa is an illustration. The first cases were identified in December of 2013; WHO was first notified of the outbreak in March of 2014 but did not declare it a public health emergency of international concern until August (Kalra et al. 2014). The response has been criticized as the result of "the combination of dysfunctional health systems, international indifference, high population mobility, local customs, densely populated capitals, and lack of trust in authorities after years of armed conflict." (Farrar and Piot 2014, p. 1545). By the time the outbreak's end was declared, over 28,000 cases had been confirmed, 40% of which were fatal. Moreover, the consequences of the epidemic reverberate. Health infrastructures have been decimated by the deaths of healthcare workers, with resulting impacts on vaccinations against diseases such as measles. Social disintegration, food insecurity, and psychological trauma remain in the epidemic's wake (Kaner and Schaack 2016). In the United States, communication missteps and subsequent mistrust led to cries for closing borders, quarantining anyone from an affected area, and augmenting powers of the federal government to restrict travel.

Gaining the information needed to respond to epidemics such as Ebola is complex. It depends on the existence of health infrastructures and trust in their use. It requires recognition of events, transmission of information about them, and analysis of the information thus gained. A failure of any of these can be devastating, as West Africa learned to its peril in 2014. Through public notice of the deliberations of the Emergency Committee, the WHO is attempting to achieve transparency and accurate communication as suggested in its 2017 surveillance guidelines (WHO 2017b). Its perceived success in meeting this goal was challenged by COVID-19, however.

2.2.3   COVID-19

COVID-19 is the disease caused by the novel coronavirus, SARS, CoV-2, which apparently emerged into human-to-human transmission in late 2019. As of just over six months into the pandemic, much was changing and remained unknown. Crystal-clear, however, was that many areas of the globe lost weeks in early 2020 that were vital to prevention of disease spread. Concerns were raised that the WHO had been too slow in recognizing the threat. Allegations that the WHO had delayed in sounding the alarm in deference to China were used by President Donald Trump to notify the WHO of the United States' intention at the time to withdraw from that organization (Rogers and Mandavilli 2020).

2.3   Reactions to Contagion

Contagious diseases like plague are nasty and scary. They are enemies that attack from outside the body. They cause distressing symptoms that are regarded with fear and distrust, such as copious bleeding, vomiting, or diarrhea. Their results may be disfigurement, odors and filth, or sudden death. When their causes are unknown, but risks of transmission appear high, the presence of contagion may lead people to avoid, stigmatize, imprison, or kill those who are thought to be sources of infection (Smith et al. 2004). Possessions, dwellings, or communities may be burned to eradicate what are thought to be sources of infection. These reactions may track, and be intensified by, lines of class, ethnicity, or especially race. They may also be linked to moral condemnation of those who are ill. Fear in the face of deadly outbreaks is neither surprising nor unjustifiable but has encouraged conducting surveillance in ways that are at best morally problematic. Here, we highlight stigma and isolation, cultural disruption, and moral condemnation as particularly serious ethical risks of even rudimentary surveillance such as counting population numbers or cases of disease.

2.3.1   Stigma and Isolation

Leprosy has been one of the most vilified diseases throughout history. To be a "leper" is to be an outcast. Caused by a bacillus, leprosy is contagious. People with leprosy have skin ulcers, lose feeling in affected areas of the body, and, in more advanced stages of the disease, experience contractures of fingers, toes, and limbs, or lose digits. They are, in short, (de)formed. Lepers have been avoided, shunned, and shunted off to far away colonies so that others would not have to see them, touch them, or risk infection from them. The *Bible* portrays leprous skin conditions as uncleanliness. Leprosy has been seen as a mark of shame from God (Grzybowski et al. 2016)—the classic "stigma."

Because individuals vary genetically in susceptibility to leprosy infection, leprosy also has been associated with particular ancestral subgroups. So, in Hawai'i, Native Hawai'ians were disproportionately infected when the disease arrived. Until seafaring European explorers reached their shores in the late eighteenth century, the Hawai'ian Islands had been isolated from contact with others and their residents had not developed resistance to many infectious diseases, including venereal disease and leprosy. Leprosy likely came to the Islands by the 1840s. By 1865 its spread had frightened authorities and the Legislative Assembly enacted "An Act to Prevent the Spread of Leprosy." The law required physicians to report all suspected cases, established a hospital, and set up an isolated colony to quarantine infected persons on the Kalaupapa peninsula on the island of Molokai.

Kalaupapa National Historical Park in Hawai'i stands today as a memorial to the 8000 people who were banished to that remote peninsula and died there of leprosy (Greene 1980). The peninsula was surrounded by the ocean on three sides and two-thousand-foot cliffs on the fourth; it could be reached only at two ocean landings and then only in good weather. Anyone suspected of leprosy was isolated on the peninsula, often by force, including some who were not ill but became so after being quarantined in close contact with others. The residents forcibly relocated to Kalaupapa did not starve—the peninsula was agriculturally fertile—but they were subjected to appalling living conditions, died at high rates, and never saw families or friends again (Tayman 2006). Father Damien, a priest who ministered to those on the colony and eventually died of leprosy himself, was sainted by the Catholic Church in 2009.

The vast majority (97%) of those isolated at Kalaupapa were Native Hawai'ians. Their isolation was imposed by the European settlers on the islands, not by the Native Hawai'ians themselves. Western attitudes of disgust towards those with leprosy were not shared by Native Hawai'ians (Amundson and Ruddle-Miyamoto 2010). The Act to Prevent the Spread of Leprosy and the policies that followed have been sharply criticized for discrimination on the basis of race and disability.

The story of Kalaupapa is gripping but not unique. Many other stories could also be told to illustrate how fear can interact with prejudice to generate disparately harsh treatment of those identified as ill who fall into disadvantaged minority groups. Chinese in San Francisco were mistakenly quarantined for plague, Haitians were stigmatized as the bearers of AIDS, and even very recently racism echoed in the panic about the possibility that Ebola would come to the United States. As Gizmodo journalist Stassa Edwards (2014) writes:

> The Western medical discourse on Africa has never been particularly subtle: the continent is often depicted as an undivided repository of degeneration. Comparing the representations of disease in Africa and in the West, you can hear the whispers of an underlying moral panic: a sense that Africa, and its bodies, are uncontainable. The discussion around Ebola

has already evoked—almost entirely from Tea Party Republicans—the explicit idea that American borders are too porous and that all manners of perceived primitiveness might infect the West.

To be sure, collection of data about births and deaths or population numbers does not by itself cause fear or stigma. Context matters: are the disease and how it spreads well understood? Can it be readily treated? Does it cause symptoms that disgust or horrify, such as uncontrollable diarrhea or hemorrhage? Does it disfigure or maim, like leprosy? Does it track—or even appear to track—racial or ethnic lines? Are those who are disproportionately afflicted from groups already disfavored for other reasons? Is the disease possibly related to conduct judged to be immoral at the time, such as prostitution? And, is there any way to regard the ill as causing their own misfortune, and thus to being responsible and blameworthy for it? If the answer to any of these questions is "yes," information about numbers may fuel fear and stigma, especially if it reveals information about population subgroups disproportionately affected by conditions that are frightening and poorly understood.

2.3.2   Cultural Disruption

The arrivals of explorers and colonizers brought globalization of disease to previously remote areas. Despite some disputes about the numbers (Roberts 1989), native populations were clearly devastated by infections such as measles or smallpox. Although at least much of the disease spread appears to have been unintentional, its impact has been compared to the Holocaust and other genocides (Brave Heart and DeBruyn 1998). At the same time, colonizers were exposed to infections novel to them but endemic in colonized areas, such as malaria or yellow fever. Catastrophic declines in indigenous population numbers, joined by environmental effects of colonization and the imposition of measures to protect colonists from the new diseases they encountered, caused extensive cultural disruption.

So-called "tropical medicine" developed intertwined with the history of colonialism, according to historian Deborah Neill's comprehensive account (2012). Emerging along with increasing understanding of the germ theory of disease, the specialty of tropical medicine was designed to protect European colonizers from new infections they encountered. It also aimed to safeguard the health of workers who were needed for development and exploitation of natural resources. While on the one hand as a specialty it contributed greatly to the understanding of diseases such as yellow fever, encephalitis, and malaria, on the other hand it often reflected racist attitudes about the superiority of Western practices of hygiene and cleanliness, and the backwardness of local populations. It also contributed to the ability of colonizers to remain in power and shaped how they exercised the power they had.

Separating population groups was the primary recommendation of tropical medicine for preventing disease spread to European colonists (Neill 2012, p. 91). Native populations were judged to be reservoirs of disease because of their poor hygiene.

Congregating them into their own defined areas was believed to reduce risks of disease transmitted by mosquitoes. These recommendations for segregation dovetailed as well with the political and economic interests of settlers. Their echoes persist today, as Neill writes (2012, p. 101): "The legacy of segregation fueled deep racial divisions that would haunt European administrations as well as emerging African nations well into the twentieth and twenty-first centuries."

Disruption caused by colonial settlements changed patterns of disease activity, just as population expansion and climate change are continuing to do today. In the late nineteenth century, for example, a particularly deadly sleeping sickness epidemic spread through central Africa, originating with changes in the habitat of the tsetse fly. In addition to surveillance, strategies adopted by European colonists to address the epidemic included forced moves of populations from infected areas, travel restrictions, and separate camps for those already ill. Those believed to be infected were subjected to treatments such as high doses of arsenic, even at levels that caused blindness and death. Not surprisingly, these impositions met with resistance and rebellion. The cultural disruption from forced relocation of entire villages was particularly extensive. Historian Helen Tilley (2016), in commentary for the American Medical Association *Journal of Ethics*, judges these efforts to have been forms of structural violence that "underpinned colonial rule" and "not only disrupted people's lives and livelihoods but also created enduring inequalities that laid the groundwork for more damage."

2.3.3 Moral Condemnation

Infectious diseases such as syphilis are sexually transmitted. Their identification and control have been associated with moral condemnation and blame of those who become infected and who transmit infection. As Chapter 3 discusses in more detail, methods of case identification and contact tracing were importantly shaped by these moralistic judgments. But even apart from such judgments about individual conduct, populations or population subgroups have been identified with the moral taint of diseases. Early on, venereal disease was associated with prostitution and lewdness; the very term "venereal" itself is rooted in the Latin for sexual love. Wet nurses were also condemned for disease transmission by staunch Calvinists seeking to "save the family from corruption" (Siena 1998). Women were generally blamed for the disease; according to Siena, beginning in the seventeenth century the theory was prevalent that venereal disease originated spontaneously from putrefaction in the womb. The U.S. has seen racist claims made about sexual licentiousness as explaining the disproportionate rates of sexually transmitted infections among blacks, although the explanations lie with social networks and access to care rather than rates of sex (CDC 2019; Adimora et al. 2006). For a time, it was believed that the natural history of diseases such as syphilis differed between blacks and whites.

Nowhere has moral condemnation of the infected been more apparent than in attitudes toward HIV/AIDS. The association of AIDS with homosexuality—it was at one point referred to as "gay related immune deficiency" or "GRID" before the virus causing it was identified—led moral conservatives and religious groups to call for criminalization of transmission of the disease. In 1988 while Ronald Reagan was president, a presidential commission advocated criminal penalties for knowing transmission of the HIV virus; approximately half the states in the U.S. enacted such laws. The U.S. Agency for International Development encouraged enactment of similar laws in sub-Saharan Africa and these laws proliferated as a result (Francis and Francis 2013). In South Africa, over 30% of women are HIV positive and these women experience high rates of intimate partner violence for their supposed infidelity even when they are pregnant and even if it is likely that they were infected by their intimate partners (Bernstein et al. 2016). Uganda, with the support of evangelical Christians, enacted an Anti-Homosexuality Act in 2014 that was struck down by its constitutional court but that continues to resonate politically.

A report by Human Rights Watch (2018) illustrates the toll that such moral panic against same sex relationships can take on groups and their health. In Indonesia in early 2016, anti-gay pronouncements were widespread in media. The country's largest Muslim organization, the Nahdlatul Ulama, urged criminalization of LGBT activism and forced reformation of LGBT people. This anti-LGBT rhetoric was followed by raids on locations including private homes where LGBT people were thought to be, with arrests, and with convictions for pornography. Human Rights Watch reports that this anti-LGBT activity has been correlated with a sharp spike in HIV infections and with increasing difficulty for public health outreach workers to reach people at risk of infection.

Such fear and condemnation of infectious disease threaten to undermine cooperation with even simple forms of surveillance. Many—but not all—of the examples we have given were associated with times before much was understood about the causes of disease or disease transmission, when avoidance and isolation seemed the only realistic responses. However, even with scientific progress, uncertainty, controversy, and reactions of mistrust remain. These must also be addressed if even simple forms of surveillance are to be reliably sustained.

2.4   Limits of Science: Risk and Uncertainty

Such reactions of fear to contagious disease were in part rooted in the limited scientific understanding of disease causation and transmission. But, however impressive advances in science may be, for epistemological and ethical reasons advances cannot be expected to resolve all problems of fear of disease. Some diseases are deservedly frightening. Science is imperfect. Analytic paradigms are contested, there is much uncertainty, and knowledge is evolving. Contested normative assessments are invoked in support of judgments about the significance of results and policy recommendations based on them. Behind these controversies lie deep epistemological cleavages. Studies of social epistemology and epistemic injustice have shed new lights on the status of scientific claims. Social epistemologists explore how social structures influence judgments about what is knowledge and what should be studied. Epistemic injustice occurs when testimony of disfavored groups is disbelieved, suppressed, or never even articulated. For example, when the testimony of women about their sexual activity is disbelieved or never heard, misunderstandings about disease transmission patterns may flourish.

Ethical problems with the conduct of science complicate these epistemological issues. Unjust exploitation of research subjects and conflicts of interest give legitimate reason for questioning scientific claims. It should thus come as no surprise that generalized suspicions of science have arisen to challenge trust in the use of science for public health initiatives. Some uncertainties in science and their impact on trust in surveillance are ineluctable. Additional problems for trust can be brought by the failure to recognize that ethical choices are intertwined with apparently simple empirical claims and, even more, by ethical malfeasance in the conduct of science.

2.4.1 Understanding Disease Etiology

One of the most—if not the most—celebrated public health successes of all time was Sir John Snow's identification of water from the Broad Street pump as the source of a cholera epidemic in London in 1854. Yet Snow's work was bitterly contested and greeted by many with disbelief.

During the mid-nineteenth century, scientists vigorously debated the causes of disease. Some attributed disease outbreaks to "miasma": bad air rising from decaying organic matter. The odors of unsanitary urban areas were thought to indicate unhealthy levels of miasma. Defenders of the miasma theory of disease argued that disease could be eradicated by proper sanitation to avoid more intense releases of miasma. These miasma theorists were important early supporters of public works and aid for the poor (UCLA Department of Epidemiology 2015). Their critics asserted the germ theory of disease: that microorganisms transmitted disease from person to person. The shift from miasma—an environmental theory of disease—to microorganisms also presaged a shift away from social programs for addressing disease to strategies singling out infected individuals.

When cholera broke out in London in 1849, Sir John Snow, an early proponent of the germ theory of disease, painstakingly correlated cholera deaths in London with the water supply. Snow's work benefited from the more systematic methods of modern public health surveillance that had been instituted in the nineteenth century. In England, the Health of Towns Committee of Liverpool appointed the first designated public health officer, Thomas Fresh, as Inspector of Nuisances in 1844 (Parkinson 2013). William Farr, superintendent of the Statistical Department of the British General Register Office from 1838 until 1880 and generally regarded as the developer of the modern concept of public health surveillance, standardized methods for collecting and analyzing vital statistics (Langmuir 1976). Edwin Chadwick, secretary of the Poor Law Commission, encouraged studies of the life and health of the London working class with the aim of improving sanitary conditions; the study recommendations for a national board of health, local district health board, and district medical officers were adopted in the Public Health Act of 1848 (Committee 1988, p. 60). Although Farr was an adherent of the miasma theory of disease, his data were useful for Snow's investigation of patterns of cholera deaths.

For several years, despite Snow's correlations, scientific controversy raged over the mechanisms of cholera spread. Then in 1854 in the Soho area of London, Snow observed the correlation between cholera deaths and drinking the water from a particular local source, the Broad Street pump. Snow urged removal of the handle from the pump and stopped the outbreak in its tracks (Hempel 2007, Chs. 14–15). With the discovery of microorganisms in the water, Snow's work provided critical evidence for the germ theory of disease.

Scientific understanding of disease causation and transmission has of course grown exponentially since Snow's day. Nonetheless, much remains unknown or controversial about disease classification and etiology. Novel infections such as HIV initially was or COVID-19 now is pose particular challenges because in the beginning little is known about the disease itself or how it may be transmitted. Lines between what are regarded as infectious conditions and what are not so regarded continue to shift. For example, cancer, once feared as infectious and transmissible, came to be regarded as an enemy solely within the body of its host. Today, however, the HPV virus is known to play a role in causing some cancers, returning these diseases into the category of conditions that can be transmitted from some to others and even prevented by vaccination. Other transmissible viruses such as Hepatitis C have also been implicated in causing some cancers. "Cancer" is coming to be regarded as a multiplicity of conditions implicating a variety of environmental, infectious, and genetic factors. Cancer research is now exploring immunotherapies that draw on paradigms from infectious disease treatment, and infectious disease treatment is likewise benefiting from developments in cancer immunotherapy, thus further blurring the lines between cancer and other disease processes (Hotchkiss and Moldawer 2014).

To take another example, studies of the human microbiome are yielding increasing knowledge of potential interconnections between varieties in bacteria and other microorganisms within the body and disease states (Lloyd-Price et al. 2017). Not incidentally for this volume, increasing understanding of the role played in human health by microbial communities in the gut, vagina, mouth, mucous membranes, and skin, may also suggest important but as yet unexplored changes in directions for surveillance. The consequences for privacy of systematic collection of material internal to the human body are significant. People from different geographic regions may have different microbiota, so it may be possible to identify their nationality from samples. Changes in the microbiome may be associated with behavior, contact with others, or prior medical treatment, too. Because microbiomes are unique to individuals, samples may provide sufficient data for individual identification, even small samples from touching human skin (Gligorov et al. 2014) Yet depending on the directions taken by scientific understanding of the human microbiome, surveillance may need to move in hitherto unexpected directions to be effective for some purposes.

2.4.2 Understanding Population Trends and Their Significance

Scientific understanding of changes in population health is also evolving. As already outlined, surveillance as a means of tracking the health of populations has a long and contentious history. Census data about population numbers can be a measure of strength—but can also be a way for enemies to discern weak spots or for those in power to learn where to extract resources. Information about disease incidence or prevalence can be both instrumental in disease prevention or a source of stigma and exclusion. Decisions to collect—or not to collect—information about populations, as well as judgments about the significance of this information—also harbor controversies with ethical implications.

The seventeenth century saw the emergence of systematic data collection about populations and the early development of epidemiology as a science. In the midst of wars and recurring outbreaks of plague, the English kings mandated the Church of England to maintain a parish-level system for recording plague deaths. The information was sent to London and published weekly as the London Bills of Mortality; those who were more fortunate were able to rely on this information to decide where to move to try to avoid the plague. Using 50 years of the Bills, John Graunt in the 1660s developed the first known tables of health data and techniques of analysis that enabled him to hypothesize fluctuating environmental causes for plague deaths. During the eighteenth century, Johann Peter Frank in Germany advocated surveillance of school health, injury prevention, maternal and child health, and public water and sewage treatment as part of a system of police medicine. Mirabeau and other French Revolutionary leaders saw population health as the state's responsibility. In the American colonies, Rhode Island required innkeepers to report contagious diseases and later enacted required reporting of smallpox, yellow fever, and cholera (Declich and Carter 1994).

By the late eighteenth century, population growth had appeared as a new threat. Thomas Malthus' argument that geometrical population growth would soon outstrip food supplies spurred Parliament to pass the Census Act in 1800. The first systematic count of the population of both England and Wales was held just afterwards, in 1801 (Office for National Statistics 2015). In the United States, the first formal census count occurred in 1790 and divided the population into free white males (over and under 16), white females, other free persons, and slaves (United States Census Bureau 2015).

To this day, data collected through the census remain a critical source for public health. During the nineteenth century, in both the United Kingdom and the US, public health efforts were often aligned with reform movements to address the conditions of the urban poor (Fairchild et al. 2010). U.S. census data provide key background information about population trends, housing characteristics, educational attainment, and economic characteristics of geographic areas divided by ZIP code. These data are combined with other data, such as health surveys, to produce community health dashboards. These compilations of health indicators can be used by communities for their own improvement, by local public health authorities, by health care providers making decisions about where to locate or what kinds of care to provide, and by businesses seeking to understand workforce availability in decisions about where to locate. They are widely used by researchers interested in studying the social determinants of health.

Despite its great utility, in the U.S. at least there also are significant barriers to many uses of census data, particularly in forms that could allow individuals to be identified. Legal restrictions on the data that may be shared, including restrictions among sharing by agencies of the federal government, are designed to protect the integrity of the data-gathering process. Under U.S. law, for example, the Census Bureau may not share identifiable information with the U.S. Department of Homeland Security.

Nonetheless, despite these protections, immigrants and others with fears about their status or about stigmatization may be especially reluctant to answer the calls of census takers. Concerns are also raised about individual privacy and the potential for targeting, especially if census data are released with narrow geographic delineations even when they have been deidentified. In response, the Census Bureau takes a number of steps to protect privacy (Census Bureau 2019). But controversies about what data the census should collect and what uses may be made of this data continue to rage in the U.S., as the Trump Administration's failed efforts to include a citizenship question on the 2020 census illustrate.

Indeed, as the case of leprosy in Hawai'i illustrates, public health interventions have not always been benign. To take another notorious example, Dr. Walter Lindley, officer of the fledgling health department in Los Angeles in 1879, called for the improvement of sanitation, the construction of a municipal sewer system—and the eradication of Chinatown. Over the decades, city health officials portrayed Los Angeles as "pristine," but its immigrant sections—Chinese, Japanese, or Mexican— as "rotten spots" (Molina 2006, p. 1). To take a much simpler current example, information about demographic trends in a community such as declining birth rates or increasing percentages of the elderly in the population, coupled with rising health care costs, declining levels of educational attainment, or poor housing stock, may lead businesses to avoid investing in that community, thus exacerbating any existing downward spiral.

2.4.3   Flaws and Gaps in the Data

From the perspective of epidemiology as a science, initial problems lie with the data itself. Some data may simply be absent, for example if population subgroups refuse to answer questions or cannot be found. Surveys using landline telephones have become increasingly unreliable as indicators of population trends, for example, because they tend to under-represent minorities and people in younger age cohorts. Researchers using these data have devised various methods of correcting for bias in survey responses, but the success of these methods remains controversial.

In addition, questions may be framed in ways that are problematic and elide or confuse important information. For example, the U.S. government categories for race, used by law to collect the census, include only these options: white, black or African-American, American Indian or Alaska Native, Asian, and Native Hawaiian or Other Pacific Island. People are asked to self-identify and, since 2000, have been given the choice to identify as more than one race (Census Bureau 2018). By relying on self-identification, the methodology captures at best perceived race and must not be understood to be measuring race in any other sense. The methodology also has been criticized for the failure to make important distinctions, for example by lumping "Asians" together into a single category.

Beyond gaps and confusions in the data lie controversies over analytic techniques. The learning algorithms developed through artificial intelligence may magnify existing deficits in the data. If people in a given location show up initially overrepresented in data about the presence of a condition, and then as a result are subject to disproportionate testing recommendations, the disproportion with which they are identified as having the condition may increase. An example is drug testing within particular subpopulations judged to have higher proportions of drug use to begin with. Testing rates in these subpopulations may simply replicate earlier suspicions. Meanwhile, the failure to test other subpopulations may miss information about even higher drug use percentages.

2.4.4 False Positives and False Negatives

When even relatively straightforward findings derived from population data are used to draw public policy recommendations, choices must be made about the comparative importance of false positive or false negative results. Take the death penalty as a harsh example. Assuming that the criminal justice system is sometimes imperfect in convicting someone of a capital offense, which is worse: putting an actually innocent person to death (false positive) or letting an actually guilty person go free (false negative)? Policymakers differ in how they answer this question about the death penalty. Some argue that the state should avoid executing the innocent at almost any cost. Others argue that public safety and the potential for deterrence matter more, even if there may be risks of executing the innocent. These policymakers may also disagree about the extent of the risks—death penalty critics tend to emphasize probabilities not only of actual innocence of the act altogether (wrong person cases) but also of innocence of an offense of the degree of severity charged (right person, wrong offense cases)—but the point here is that they disagree about how to weigh ethically the respective probabilities of these errors. And they disagree about where the burden of persuasion should lie when information is incomplete: is it up to a proponent of the death penalty to argue that it deters, or up to an opponent to produce evidence that it does not?

Similar judgments are made all the time about medical tests. If detecting a positive is very important, because it might enable a lifesaving intervention, thresholds of suspicion arguably should be set high. The sensitivity of a test—whether it misses actual positives—will be judged more important than its specificity, whether it also sweeps up a comparatively high number of false positives. On the other hand, if the risks of over-identification are high, as they would be if there is a significant probability of executing people who are actually innocent, arguably the specificity of a test—whether it identifies a high number of positive cases that are false positives—will be judged more important than its sensitivity. An underlying complication to setting these thresholds is that in populations in which true positives are infrequent, the probability that a positive test is a true positive is low, whereas in high frequency populations it is higher. Failure to understand this complication is known as the base rate fallacy. If rates of false positives and false negatives are not well understood, the result may be policies that are based on erroneous assumptions about either the frequency, or the comparative absence, of a condition of concern.

Value judgments are part of the equation in deciding how to set these thresholds. People might differ about how to weigh the risks and costs of intervening with people who test positive and the importance of the goal of the intervention itself. These differences might apply to judgments about the goals of intervention for the individual him or herself, or about the importance of protecting others from the individual. People might also differ about the extent it is reasonable to defer to expertise and whether it is morally acceptable to intervene paternalistically with people for their own good. Some might argue that even if individuals are likely to make poorly reasoned judgments because they lack understanding of simple points about probability such as the base rate fallacy, these decisions should still be theirs to make and they should be allowed to take risks at least with themselves, if not to protect others from them. There may also be differences of opinion about how much an individual's decision affects only him or herself; for example, in the controversy about whether to require vaccination there are disagreements about whether its primary goal is protection of the individual or protecting others from the individual's becoming a source of contagion. A further complication is that there might be reasons for questioning expert advice. Conflicts of interest—well known for their influence on medical practice—are one ground for why expert advice might come under fire. Injustice—or the history of injustice—is another, as we discuss further in a later section of this chapter.

Screening mammography for breast cancer is a particularly controversial example of these problems with the sensitivity and specificity of methods for screening populations (Plutynski 2012). On the one hand, early detection of these cancers is thought to be important for both morbidity and longer-term mortality. If so, some argue, the threshold for suspicion should be set high. Moreover, defenders of such a high threshold argue, the worry and subsequent evaluation associated with assessing whether an initially identified positive is a true positive should be judged far less serious than the risks of an early death from cancer. On the other hand, setting a high threshold may recommend invasive and potentially risky treatment for women who would never have needed it because their early cellular changes might never have become cancerous. Further, critics of a high sensitivity threshold say, the impact of worry and repeat testing should not be so readily discounted. One unanticipated consequence of undergoing evaluation for a false positive may be to discourage the willingness to return for mammography screening in later years, thus decreasing the likelihood that women once identified as false positives will return for medically appropriate care (Shen et al. 2018). Concerns about medical paternalism loom large in these debates. So as well do allegations of conflicts of interest on the part of providers who may have economic incentives to over-test and over-treat.

Public health surveillance likewise raises these normative questions about the import of false positives and false negatives. Signals from reports of mortality or morbidity also may be misread. Consider an apparent uptick in reported influenza cases requiring hospitalization or resulting in death. This uptick might be an accurate signal of the spread of a very serious form of influenza—a true positive signal. Or, it might be a flaw in the data if cases are counted twice because they were reported by both the initial diagnosing physician and the treating hospital. Or, it might be the result of misdiagnoses of a respiratory infection, or of selective case reporting—both misleading as signals of an influenza outbreak. If only the very sick seek out hospital care, while many others who were infected simply treat themselves at home, reports of fatality rates based on the hospital data may be greatly exaggerated. On the other hand, there may be underestimations based on available data if patients never come to hospitals, reports of their conditions are not made, or tests erroneously report that they are not ill. These judgments and the policy recommendations based on them will be affected by value judgments and problems that may be raised about them.

To illustrate, in 2009 reports surfaced of an outbreak of a strain of influenza in Mexico that appeared to be very serious. Reports were that the outbreak had reached epidemic status and would soon become pandemic. World-wide panic ensued, with school closures, travel bans, and widescale rejection of Mexican products even though they could not be sources of contagion. The WHO declared a public health emergency of international concern. Because the virus had originated from pork, the flu was called "swine flu." Pork producers were hit particularly hard even though the disease could not be spread by eating pork. The economic consequences for Mexico in particular were extensive: one estimate put tourist industry losses at $2.8 billion overall and pork production losses at $2 million monthly (Rassy and Smith 2013). Clearly, the decision to declare a public health emergency had major economic and social consequences for Mexico. Whether it was truly such an emergency, however, was another question.

The 2009 outbreak was the first test in the context of a potential pandemic of the WHO International Health Regulations that had entered into force in 2005. After the supposed pandemic proved "a damp squib," the WHO was resoundingly criticized for overreacting to the event and for undisclosed conflicts of interest that had allegedly affected the decisions it made (Godlee 2010). Estimates of the severity of the form of influenza were one of the problems: because only those who were seriously ill had appeared for treatment in the area of Mexico where transmission of the disease was thought to have started, and these patients had high rates of mortality and morbidity, unwarranted conclusions were quickly drawn about the lethality of the influenza strain. A subsequent WHO report responding to the criticism emphasized that "the main ethos of public health is one of prevention: ... in the fact of uncertainty and potentially serious harm, it is better to err on the side of safety" (WHO 2011). That several WHO scientific advisors had not disclosed their economic ties to pharmaceutical companies manufacturing influenza prophylactics did not help trust in these judgments, however. While finding "no evidence of malfeasance," the report did acknowledge the uncertainties involved in the decisions that WHO had made and the importance for all countries of making decisions such as imposing travel restrictions on the basis of sound evidence (WHO 2011, p. 11). By contrast, as we described above, the WHO came under criticism for delaying judgments about the significance of the Ebola epidemic in West Africa in 2014. And it is subject to a chorus of criticism about alleged delays in reporting the COVID-19 situation and in declaring it a Public Health Emergency of International Concern. Indeed, President Trump cited the WHO's supposed failures and favoritism to the Chinese to justify his decision to give notice that the U.S. intended to withdraw from the WHO (Rogers and Mandavilli 2020).

Science and decisions based on it take place under conditions of admitted uncertainty. Judgment calls are made about data collection, analysis, and subsequent policy recommendations. If judgments are not carefully made and carefully explained—and this may be difficult in the face of panic—the result may be serious and unwarranted disadvantage for some. Such results may be unjust or complicate existing injustice, particularly if protective judgments benefit those who are already better off or entrench structural injustice. They may also undermine trust in data, generating reluctance to participate in or believe the pronouncements of subsequent surveillance activities.

2.4.5 *Behavioral Economics, Cognitive Biases and Judgments of Risk*

A further complication in these reactions to threats of disease is that human beings are not very good at understanding risks, even when they have been professionally trained to do so. Recent empirical work about how people make decisions has documented common cognitive biases that are relevant to risk judgments. These biases are useful shortcuts to reasoning but they are also frequently misleading. Understanding these biases and how they may affect risk judgments is critical to understanding what is to be surveilled and how the results of surveillance may be communicated and understood.

One such bias is called the "availability heuristic." Famously identified by Tversky and Kahneman (1973), this bias describes how people are likely to judge that a risk is greater when they have recently heard about a relevant occurrence. For example, after a highly-publicized airplane crash, people rate air travel as less safe than they otherwise would. During the Ebola epidemic, people believed that the chances of transmission in the U.S. from someone returning from the affected area were far higher than they actually were. Such misjudgments occurred not only in the general population but even among professionals; in the U.K., for example, diagnoses of possible viral hemorrhagic fever rose markedly when the Ebola epidemic was prominent in media reports (Curran 2015). In clinical care, health care providers are disproportionately likely to identify a patient as having a particular condition when they have just experienced missing a diagnosis in another case that became memorable for them because of the harm that resulted to a particular patient.

Relatedly, people may be likely to judge that a particular case is representative of a population when it is seen as similar to other population members, even if it is not. The judgment that the severe cases of flu seen in Mexico in the 2009 epidemic were representative exemplified this bias. Outbreak reports of cases that are not randomly selected are notably prone to this bias of assuming that they are representative of the population more generally (Curran 2015).

Anchoring and framing are another type of relevant cognitive bias in judgments of risk. With this bias, how a problem is defined may affect the judgments that are made. Well-known examples of this bias are whether an outcome is presented in frequency or percentage form and whether it is presented in terms of possibility of death or possibility of survival. That is, people will judge an outcome as more likely if they are told that it occurs in 10 out of 100 cases than if they are told it occurs 10% of the time. And people are more likely to choose an alternative with a 95% chance of survival than with a 5% chance of death—even though numerically speaking the two are equivalent. Framing in terms of gender, race or ethnicity, age or disability may also affect clinical judgment in ways that lead to errors (Bui et al. 2015).

Judgments of immorality complicate these cognitive biases. Recent empirical work on risk perception indicates that people tend to believe that risks are greater when they judge conduct to be immoral. For example, when parents leave their children for short periods of time for reasons people judge to be morally problematic, people judge risks to the children as higher than if the reasons are seen as innocent (Thomas et al. 2016). Applied to disease, this would suggest judgments of risks of transmission are higher when conduct is judged to be immoral. This phenomenon might explain why prostitutes are judged more likely to transmit disease than clients who visit them. When actions are judged to be intentional, moreover, any harm they cause is seen as greater than if the same results occurred from actions not judged to be intentional (Ames and Fiske 2015). And causation is more likely to be attributed to actions judged to be immoral (Alicke 1992).

How these cognitive biases affect public perceptions of risk has been the subject of extensive study. From plague to Ebola, the threat of disease has been both exaggerated and discounted through these biases. Their impact on clinical care has also been studied but primarily with hypothetical rather than real-world examples (Saposnik et al. 2016; Blumenthal-Barby and Krieger 2015). People are more likely to see risks as serious when these heuristics are at work. Heuristics may mount up, too, and thus may play a role in why presuppositions about populations are so difficult to shake. It might be surmised that public health professionals are no exception in the demonstration of cognitive biases, although this apparently has not been systematically examined to any extent (Greenland 2017).

2.5   Suspicions of Science: Exploitation of Research Subjects and Conflicts of Interest

Uncertainties about scientific judgments have been compounded by the impact of scientific misconduct such as exploitation or racism. Public health researchers have exploited research subjects in ways that reveal clear racism. Public health recommendations also have been met with complaints about conflicts of interest that have grounded mistrust. While the impact of misconduct should not be exaggerated, for trust in data collection and use to be sustained, these concerns must be acknowledged and addressed.

2.5.1   Exploitation in Research

The legacy of the Tuskegee syphilis study casts a long shadow over attitudes towards public health and medicine in the United States today. But it is not the only shadow; eugenics, racism, and the exploitation of human participants in research have all played roles, not only in Nazi Germany but in many other circumstances. The Tuskegee study has become iconic in explaining reluctance among Blacks in the U.S. to trust health care institutions, although its role has perhaps been distorted in comparison to other factors. The study has been particularly well described by the feminist historian Susan Reverby (2009) whose excellent account is the basis for some of the discussion that follows.

   In the early part of the twentieth century, syphilis was feared for how it ravaged both body and mind. Because the disease was known to be sexually transmitted, those who were infected were often stigmatized and morally condemned for their supposed licentiousness. Prevalence of the disease was little understood and only primitive treatment in the form of mercury and arsenic was available. In the U.S., the disease was thought to be a particular problem among blacks because of what was supposed to be their sexual irresponsibility. Syphilis also was judged to take different forms in blacks and in whites, supposedly for genetic reasons—and there were links between anti-syphilis campaigns and eugenics. By the 1930s, syphilis had become a subject of major public health concern and a particular target of the U.S. Public Health Service; Chapter 3 describes the development of case identification and contact tracing as a response. Before effective treatment became available, however, the natural history of the disease was of particular scientific interest. Macon County, Alabama, was selected for study because of its high black population, high incidence of disease, and presence of the respected Tuskegee Institute for black education. People in the county were poor, illiterate, and lacked access to health care. Several demonstration projects for syphilis treatment had been instituted in the effort to improve the health of blacks in the south, including one in Macon County, but largely without success. After the demonstration projects ended, the "Tuskegee Study of Untreated Syphilis in the Negro Male" began in 1932 to learn more about the natural history of the disease. The Study was justified by those who initiated and continued it by the importance of addressing scientific uncertainties about the course of the disease. When the Study started, the only treatment available, neosalvarsan, was risky, expensive, and required frequent injections that were difficult to provide in a rural setting. By the late 1940s, penicillin therapy had become available, yet the Study continued until 1972—another nearly 25 years. (Parenthetically, although responsibility for conceptualizing the Study has traditionally been attributed to a minor public health official, recent revelations indicate that the later head of the Public Health Service who campaigned fiercely against syphilis, Dr. Thomas Parran, may have played a role in its origins. As a result, the University of Pittsburgh has decided to remove Parran's name from the building that houses its school of public health, over which Parran served as dean.)

The Tuskegee Study included six hundred Black males, 399 with syphilis and 201 controls. Its aim was to observe the natural course of untreated syphilis. Participants entered the Study with a promise of $50 for burial expenses. They were deceived into believing that they were receiving treatment for their "bad blood," but they were not receiving any treatment at all. Instead, they were knowingly kept in the Study for long after penicillin had become readily available as an effective treatment for syphilis. Their sexual partners and their children were also at risk from their untreated infections, although syphilis is not contagious in its late stage. A reporter's revelations about the Study in the *Boston Globe* finally resulted in its termination in 1972. In 1997, survivors of the Study and relatives who had been injured received a formal apology from President Clinton. Survivors also sued and received compensation; the Trump administration has contended that any leftover settlement funds should revert to the U.S. government but descendants of the participants object (Reeves 2017).

"Tuskegee" has become a metaphor for public health abuse in the U.S. It created momentum for development of the regime for regulating the ethics of research in the United States, a regime that primarily emphasizes the individual informed consent that was judged to have been so clearly lacking in the Tuskegee Study—and that we raised questions about for public health in Chapter 6. In Reverby's judgment, the frequent comparison in ethics discussions between Tuskegee and Nazi medical experimentation mistakenly fixes it in the 1930s rather than as the ongoing racism that it continued to demonstrate. It thus constructs Tuskegee more as a violation of individual autonomy than as the manifestation of structural injustice that it actually was. Mistakenly understood in comparison to Nazi racism, Reverby contends (2009, p. 187), Tuskegee became a convenient explanation for minority mistrust of health care and medical research and its condemnation has become a code for expressing supposed racial sensitivity. Instead, the Study should serve as a reminder of the many social factors that contribute to racial disparities in health today in the U.S. Reverby thus concludes (p. 194), "The arguments about the problems of informed consent and the development of the entire institutional review board infrastructure [required by the federal regulations for the conduct of research with human subjects] provided little focus on the links between race and science or the problems of equity."

As AIDS appeared on the scene, taking its toll disproportionately on Blacks, conspiracy theories about the role of medical experimenters in transmitting the disease circulated widely. Reverby relates (p. 200) the spread of false rumors that federal government health workers had deliberately infected Tuskegee participants with syphilis, thus perpetrating American genocide. Similar rumors also circulated about AIDS, both in the US and across the globe, generating ongoing mistrust of scientists especially those with connections to governments. These rumors have continued to be used in opposition to vaccination or other medical interventions. Tuskegee and the cultural meanings it later assumed are a powerful illustration of how injustice in the conduct of science can create legacies of mistrust for public health data collection and its use to address health problems.

2.5.2   Conflicts of Interest

Contemporary health care and research, both in the U.S. and elsewhere, are beset by economic incentives. Pharmaceutical companies are a major part of the mix, but so also are providers, insurers, and researchers. Public health and even charities furthering public health goals have also come under suspicion of the taint of conflicts of interest affecting the decisions they make. Evidence is robust that people are influenced by their economic interests even when they sincerely believe that they are not.

The role of for-profit pharmaceutical companies in influencing science and, in turn, public health, is extensive. Although most major medical journals require disclosures of conflicts of interest, pharmaceutical companies fund a great deal of medical research. Funding influences decisions about what to study and what research questions to ask. It influences which end points are chosen as measures of success such as whether an average of two months increased survival is a meaningful result in comparison with the costs and side effects of a novel therapeutic agent. Fleck and Danis (2017) point out the complex normative judgments involved in making recommendations about whether such a therapeutic agent should be cleared for marketing or paid for out of shared funds, especially when trials of the drug may not have included measures of the quality of life of patients taking it or its efficacy in comparison with other treatments. They argue that when prices are very high—as high as $750,000 for limited gain such as a two month increase in survival—it is ethical to refuse to pay for the treatment out of shared funds and to expect patients who wish this option to purchase an insurance rider to pay for it.

Pharmaceutical company sponsorship also influences decisions about how to structure clinical trials, for example whether to test a new agent against placebo or against a therapy currently in use. Showing that the new agent has results comparable to the current therapy may not indicate that it is effective but could only show that both are similarly ineffective. Drug company sponsorship also influences whether replicability or reproducibility of studies are conducted. Despite requirements in the U.S. that clinical trials be registered with a federal data base,

ClinicalTrials.gov, negative study findings may not be published. Findings may never become public that a particular drug had no effect on the condition of interest. One study of clinical trials involving children found that 30% did not result in publication and that the rate of non-publication for industry-sponsored trials was double that of trials sponsored by academic institutions (Pica and Bourgeois 2016).

Importantly for public health, conflicts of interest may also affect judgments about when conditions should be considered pathological and in need of intervention. Identification of hypertension is an example. The American Academy of Family Physicians and the American College of Physicians declined to endorse American Heart Association and the American College of Cardiology guidelines identifying hypertension as blood pressure over 130/80, a figure that would classify 46% of U.S. adults as in need of intervention, in part out of concerns over conflicts of interest (AAFP 2017). One commentator has characterized this dispute not just as a conflict between specialists and generalists, but as "instead about medical care (primary *or* specialty care) versus public health. From the public health perspective it is inevitably a losing proposition and a rearguard action for doctors to treat mildly elevated blood pressures with medicine or even individual lifestyle advice" (Husten 2018).

Earlier in this chapter we pointed out concerns about conflicts of interest in the World Health Organization's decision to declare the 2009 influenza outbreak a health emergency of international concern. This declaration was the initial test of the new World Health Regulations that had entered into force in 2007. These regulations allowed for the declaration of public health emergencies of international concern; and, unlike the prior regulations they replaced, they did not limit such declarations to a specific list of serious contagious diseases such as cholera or plague. Under Article 48 of the Regulations, the Director of WHO was to appoint an Emergency Committee of experts to advise on whether an outbreak constituted an emergency of international concern. The experts were to be selected from the WHO roster of experts for advisory panels. Membership of the Emergency Committee responsible for declaring pandemic status was not made public, a lack of transparency that gave rise to concerns that these members may have been influenced by industry ties (Godlee 2010). Reportedly, pharmaceutical companies made high profits—estimates ranged from $7 billion to $10 billion, depending on the company—from the sale of flu vaccines (Godlee 2010). In addition, the expert responsible for authoring WHO's guidance on the use of antivirals during a pandemic reportedly was receiving payments at the same time from Roche, maker of Tamiflu, the principal antiviral being stockpiled for prophylactic use in case of an outbreak (Cohen and Carter 2010). The Council of Europe Parliamentary Assembly was sharply critical of the WHO decision to declare a pandemic and called for more transparency in making these decisions (Council of Europe 2010). Although as we described above the WHO's subsequent assessment of the declaration concluded that there had been no malfeasance, suspicion lingered. The report and the heat it generated indicate the importance of fully disclosing and appropriately managing the impact of conflicts of interest for trust in the science used by public health. As

WHO establishes partnerships with non-state actors in public health, conflicts of interest are a principal concern, as we discuss further in Chapter 7.

Another illustration of how known financial incentives can undercut trust in public health recommendations is the campaign by Merck in support of Gardasil, the vaccine it manufactured against the human papilloma virus (HPV) (Schwartz et al. 2007). The HPV virus is sexually transmitted and can cause a variety of cancers, primarily of the cervix, penis, and tongue. Development of a vaccine against the infection was regarded as a highly important public health measure especially for areas of the globe where regular access to cervical cancer screening was not available. The vaccine was not cheap: to the contrary, at $120 per each of the required three doses, it was the most expensive vaccine ever marketed. In the media, it was eye-catchingly described as a vaccine that could prevent cancer. After U.S. licensure of the vaccine in 2006, the CDC Advisory Committee on Immunization Practices recommended that all women ages 13–26 receive the vaccine. Nearly half of the states considered legislation requiring vaccination for school entry. When Merck's extensive behind the scenes lobbying to promote the vaccine became known, controversy erupted over the recommendations for its routine administration. To be sure, the vaccine requirements were controversial for other reasons as well: it protected against a sexually transmitted disease, the recommendations applied to young teenagers not believed to be (or desired to be) sexually active, the recommendations were perceived as interfering with parental authority over their children, and the recommendations applied to females only (despite the fact that males could transmit the disease and were themselves at some risk from the infection). The controversy lingered and fed into more general suspicions of immunization prevalent in the U.S. and elsewhere at the time (Colgrove et al. 2010).

2.6   Suspicions of Science: Skepticism and Politics

The ground is thus fertile for suspicion of science. Trust or distrust in science may be shaped by stages of a health threat and perceptions of its severity among the population. Uncertainties and contested normative judgments have been complicated by serious ethical problems in the conduct of science such as exploitation of research subjects or conflicts of interest. In addition, science has become political, figuring in debates over clean water and air, climate change, abortion, and many other hot button issues. Distrust of science is widespread, as the case of vaccination illustrates.

Vaccines to prevent disease are one of the most powerful weapons public health can deploy. Yet suspicions about their efficacy and risks—that is, about vaccine science—continue to depress vaccination rates. Maintaining the levels of vaccination in the population needed to achieve herd immunity—that level of population immunization needed to prevent disease spread—remains a fragile enterprise. Self-interested objections to vaccination as expecting individuals to take risks for the benefit of others in the community is part of the problem. So are parents objecting to taking any risks on behalf of their children for the benefit of others. But behind these objections lie mistrust of scientific representations of the efficacy, benefits, and risks of vaccination.

In 1998, a paper published in the UK in the highly regarded medical journal *The Lancet*, purported to establish a link between vaccination against measles, mumps, and rubella (MMR), and autism. The paper, by Wakefield and 12 colleagues (1998), reported a series of 12 cases in which children who had been immunized later were allegedly found to have pervasive developmental disorders. The widely publicized paper caused a sensation that resulted in vaccination refusals to the point that herd immunity was compromised in areas of the UK and elsewhere. From the beginning, a red flag was that the report was highly speculative in generalizing from twelve cases; and epidemiological studies soon questioned the findings (Rao and Andrade 2011). Six years later, ten of the twelve original coauthors retracted their support for the claims about causal relationships between immunization and developmental disorders made in the article (Murch et al. 2004). Although an initial investigation by *The Lancet* found no misconduct in the publication, it did report that Wakefield had failed to report as a conflict of interest that he had been funded by the Legal Aid Board for a pilot study that also included some of the same children, funding that could have been perceived as a conflict of interest that Wakefield should have revealed (*Lancet* 2004). Finally, in 2010, *The Lancet* fully retracted the Wakefield paper based on conclusions that the children reported in the series had not been consecutively referred and the study had not received local ethics committee approval (*Lancet* 2010).

Despite full scientific repudiation of the supposed link between vaccination and autism, suppositions about the connection will not die out. Parents, worried about the impact of the many vaccines now given their infants, continue to report beliefs about sudden behavior changes after immunization. "Anti-vax" has become a movement world-wide and vaccination rates have again slipped below herd immunity levels in some locations. The result has been outbreaks of disease, including one from exposures at the Disneyland resort in California that sickened more than 125 children and resulted in legislation repealing California's "personal belief" exemption from vaccination requirements for school children.

Suspicions about vaccine safety are one explanation for low immunization rates. Poverty and sporadic access to medical care are others. It is notable, however, that vaccine mistrust is not always correlated with poverty or lack of education; in California, for example, wealthy Marin County had one of the highest rates of unvaccinated children at the time of the Disneyland outbreak. Many parents opposed to vaccination are also suspicious of the role of the state more generally. There are high rates of vaccine refusal in families who home school their children and the California statute requiring proof of immunization for school entry exempts children who are home schooled (California 2020). Home schooled children of course do not stay at home all of the time and may be out and around places where children gather—such as Disneyland—just as much as other children are. Vaccine refusal has reportedly also been taken up as a cause by women who had adverse experiences with medicalized birth such as unanticipated caesarean section deliveries and who see vaccination as "unnatural" (Reich 2016). These women may feel that their understanding of the birthing process was ignored by physicians—put in more philosophical terms, that epistemological injustice deprived them of an opportunity for natural childbirth that they treasured.

Vaccine denialists sometimes draw on alternative theories of knowledge in support of their views. Mark Navin (2013) relates how leaders of the anti-vaccination movement describe their experiences with physicians who failed to listen to their concerns about their children. According to Navin, these denialists see themselves as victimized by epistemological injustice of disbelief in their testimony. They claim a kind of "democratic" authority as citizen-scientists who have been ignored by experts. These arguments gain credence, of course, if the supposed experts can be discounted for other reasons, such as the conflicts of interest recounted above. Despite his own conflicts, Wakefield is regarded in the anti-vax world as a hero who listened to and believed his patients and who has been unjustly vilified by a scientific establishment that may itself be corrupt. Navin's conclusion is not that the vaccine denialists are justified—he describes in detail their own epistemic vices—but that public health advocates must not only educate about the scientific case in favor of vaccination but also attend to their own epistemic vices such as the failure to listen to concerned parents who may have made accurate observations about their children.

The story of autism fear and vaccines has drawn particular attention for its lessons for scientific communication. At the time the fraudulent paper was published, worries were already running high about increasing, but unexplained, rates of autism in the population. Wakefield's paper struck a sympathetic chord among parents who worried about toxic combinations injected into their children. According to communication scholars Burgess et al. (2006), apparently causing autism had a high "outrage" factor among the public because it was judged to have inflicted severe neurological damage on innocent young children. Parents, moreover, were regarded as having been wrongfully coerced by the British National Health Service's reimbursement policies favoring the combined form of MMR immunization. Scientists, by contrast, emphasized the low or absent probability of the connection, but these claims did not have the same popular salience. Advisory committees recommending policy decisions such as the Advisory Committee on Immunization Practices that makes vaccine recommendations to the US CDC must manage these fears (Martinez 2012). In Chapter 7, we consider further how communication through groups may be helpful in achieving the values of respect for individuals that have led some to insist that parental decisions not to consent to their children's vaccination should be honored without further question.

2.7 Summary

Vital statistics can provide important information about the health of populations. For centuries, together with reports about outbreaks, acquiring this information was the primary way for people who were more fortunate to avoid exposure to diseases that could not otherwise be treated. Those who were not so fortunate succumbed. Today, with increasing understanding of disease and availability of prevention or treatment, the advantages of outbreak detection may be shared far more widely and more equally. Nonetheless, outbreak detection can generate fear and hostility if patterns of disease track otherwise disfavored groups. On the other hand, COVID-19 has revealed the importance of demographic data about the distribution of disease burdens—data that may either generate mistrust as people see their disadvantage starkly, or that may foster trust if the result is increased attention to disparities in treatment and in health.

The advance of science, particularly in understanding causation and treatment of contagious disease, made it all the more compelling to identify outbreaks quickly. But scientific advancement is not a panacea. Scientific uncertainty is ineluctable and scientific knowledge is incomplete. When there is reason to believe the practice of science cannot be trusted—because of conflicts of interest, politicization, outright fraud, or exploitation—the consequences may be dire for public health activities using even simple forms of data such as vital statistics. Trust in science to deal with what can be learned from surveillance requires not only scientific integrity but also the appearance of scientific integrity and successful communication of this appearance.

References

Adimora, Adaora A., Victor J. Schoenbach, and Irene A. Doherty. 2006. HIV and African Americans in the Southern United States: Sexual Networks and Social Context. *Sexually Transmitted Diseases* 33 (7): S39–S45.

Alicke, Mark D. 1992. Culpable Causation. *Journal of Personality and Social Psychology* 63 (3): 369–378.

American Academy of Family Physicians (AAFP). 2017. AAFP Decides to Not Endorse AHA/ ACC Hypertension Guideline. *AAFP News* (Dec. 12) [online]. https://www.aafp.org/news/ health-of-the-public/20171212notendorseaha-accgdlne.html. Accessed 12 July 2020.

Ames, Daniel L., and Susan T. Fiske. 2015. Perceived Intent Motivates People to Magnify Observed Harms. *Proceedings of the National Academy of Sciences* 112 (12): 3599–3605.

Amundson, Ron, and Akira Oakaokalani Ruddle-Miyamoto. 2010. A Wholesome Horror: The Stigmas of Leprosy in 19th Century Hawaii. *Disability Studies Quarterly* 30: 3–4.

Bernstein, Molly, Tamsin Phillips, Allison Zerbe, James A. McIntyre, Kirsty Brittain, Greg Petro, Elaine J. Abrams, and Landon Myer. 2016. Intimate Partner Violence Experiences by HIV-Infected Pregnant Women in South Africa: A Cross-Sectional Study. *BMJ Open* 6: e011999. https://doi.org/10.1136/bmjopen-2016-011999.

Blumenthal-Barby, J.S., and Heather Krieger. 2015. Cognitive Biases and Heuristics in Medical Decision Making: A Critical Review Using a Systematic Search Strategy. *Medical Decisionmaking* 35 (4): 539–557.

Brave Heart, Maria Yellow Horse, and Lemyra M. DeBruyn. 1998. The American Indian Holocaust: Healing Historical Unresolved Grief. *American Indian and Alaska Native Mental Health Research* 8 (2): 60–82.

Bui, Thanh C., Heather A. Krieger, and Jennifer S. Blumenthal-Barby. 2015. Framing Effects on Physicians' Judgment and Decision Making. *Psychological Reports: Mental & Physical Health* 117 (2): 508–522.

Burgess, David C., Margaret A. Burgess, and Julie Leask. 2006. The MMR Vaccination and Autism Controversy in United Kingdom 1998–2005: Inevitable Community Outrage or a Failure of Risk Communication? *Vaccine* 24: 3921–3928.

California (Health & Safety Code § 120335(f)) (California) (2020).

Centers for Disease Control and Prevention (CDC). 2019. HIV and African Americans. https://www.cdc.gov/hiv/group/racialethnic/africanamericans/index.html. Accessed 12 July 2020.

Choi, Bernard C.K. 2012. The Past, Present, and Future of Public Health Surveillance. *Scientifica (Cairo)* 2012: 875253.

Cohen, Deborah, and Philip Carter. 2010. WHO and the Pandemic Flu "Conspiracies". *British Medical Journal* 340: c2912.

Colgrove, James, Sara Abiola, and Michelle M. Mello. 2010. HPV Vaccination Mandates—Lawmaking amid Political and Scientific Controversy. *New England Journal of Medicine* 363: 785–791.

Committee for the Study of the Future of Public Health (Committee). 1988. *The Future of Public Health*. Washington, DC: National Academy Press.

Council of Europe. 2010. The Handling of the H1N1 Pandemic: More Transparency Needed. https://assembly.coe.int/CommitteeDocs/2010/20100604_H1N1pandemic_e.pdf. Accessed 12 July 2020.

Curran, Evonne T. 2015. Outbreak Column 16: Cognitive Errors in Outbreak Decision Making. *Journal of Infection Prevention* 16 (1): 32–38.

Dean, Katherine R., Fabienne Krauer, Lars Walløe, Ole Christian Lingjærde, Barbara Bramanti, Nils Chr Stenseth, and Boris V. Schmid. 2018. Human Ectoparasites and the Spread of Plague in Europe During the Second Pandemic. *Proceedings of the National Academy of Sciences*. https://doi.org/10.1073/pnas.1715640115.

Declich, Silvia, and Anne O. Carter. 1994. Public Health Surveillance: Historical Origins, Methods and Evaluation. *Bulletin of the World Health Organization* 72 (2): 285–304.

Dunn, Halbert L. 1936. Vital Statistics Collected by the Government. *The Annals of the American Academy of Political and Social Science* 188: 340–350.

Edwards, Stassa. 2014. From Miasma to Ebola: The History of Racist Moral Panic Over Disease. *Jezebel* (Oct. 14) [online]. https://jezebel.com/from-miasma-to-ebola-the-history-of-racist-moral-panic-1645711030?utm_campaign=socialflow_jezebel_facebook&utm_source=jezebel_facebook&utm_medium=socialflow. Accessed 12 July 2020.

Fairchild, Amy L., David Rosner, James Colgrove, Ronald Bayer, and Linda P. Fried. 2010. The EXODUS of Public Health: What History Can Tell Us About the Future. *American Journal of Public Health* 100 (1): 54–63.

Farrar, Jeremy J., and Peter Piot. 2014. The Ebola Emergency–Immediate Action, Ongoing Strategy. *New England Journal of Medicine* 371: 1545–1546.

Farmer, Paul. 2020. *Fevers, Feuds, and Diamonds: Ebola and the Ravages of History*. New York: Farrar, Straus and Giroux.

Fleck, Leonard, and Marion Danis. 2017. How Should Therapeutic Decisions about Expensive Drugs Be Made in Imperfect Environments? *AMA Journal of Ethics* 19 (2): 147–156.

Francis, John G., and Leslie P. Francis. 2013. HIV Treatment as Prevention: Not an Argument for Continuing Criminalisation of HIV Transmission. *International Journal of Law in Context* 9 (4): 520–534.

Gensini, Gian Franco, Magdi H. Yacoub, and Andrea A. Conti. 2004. The Concept of Quarantine in History: From Plague to SARS. *Journal of Infection* 49 (4): 257–261.

Gligorov, Nada, Lily E. Frank, Abraham P. Schwab, and Brett Grusko. 2014. Privacy, Confidentiality, and New Ways of Knowing More. In *The Human Microbiome: Ethical, Legal and Social Concerns*, ed. Rosamond Rhodes, Nada Gligorov, and Abraham Paul Schwab, 107–127. New York: Oxford University Press.

Godlee, Fiona. 2010. Conflicts of Interest and Pandemic Flu. *British Medical Journal* 340: 1256–1257.

Grajalez, Carlos Gómez. 2013. Ancient Censuses. In Royal Society of Statistics, *Great Moments in Statistics*, p. 21. https://rss.onlinelibrary.wiley.com/doi/pdf/10.1111/j.1740-9713.2013.00706.x. Accessed 12 July 2020.

Greene, Linda W. 1980. Exile in Paradise: The Isolation of Hawai'i's Leprosy Victims and Development of Kalaupapa Settlement, 1865 to the Present. https://catalog.hathitrust.org/Record/002237677. Accessed 12 July 2020.

Greenland, Sander. 2017. Invited Commentary: The Need for Cognitive Science in Methodology. *American Journal of Epidemiology* 186 (6): 639–645.

Grzybowski, Andrzej, Jaroslaw Sak, Jakub Pawlikowski, and Malgorzata Nita. 2016. Leprosy: Social Implications from Antiquity to the Present. *Clinics in Dermatology* 34 (1): 8–10.

Hempel, Sandra. 2007. *The Strange Case of the Broad Street Pump: John Snow and the Mystery of Cholera*. Berkeley: University of California Press.

Hotchkiss, Richard S., and Lyle L. Moldawer. 2014. Parallels Between Cancer and Infectious Disease. *New England Journal of Medicine* 371: 380–383.

Human Rights Watch. 2018. "Scared in Public and Now No Privacy": Human Rights and Public Health Impacts of Indonesia's Anti-LGBT Moral Panic (July 1). https://www.ecoi.net/en/document/1437353.html. Accessed 12 July 2020.

Husten, Larry. 2018. The Blood Pressure Guideline War Is Not A Fake War. *Forbes* (Feb. 11) [online]. https://www.forbes.com/sites/larryhusten/2018/02/11/the-blood-pressure-guideline-war-is-not-a-fake-war/#5825631d2cfe. Accessed 12 July 2020.

Jew Ho v. Williamson, 103 Fed Rep. 10 (1900).

Kalra, Sarathi, Dhanashree Kelkar, Sagar C. Galwankar, Thomas J. Papadimos, Stanislaw P. Stawicki, Bonnie Arguilla, Brian A. Hoey, Richard P. Sharpe, Donna Sabol, and Jeffrey A. Jahre. 2014. The Emergence of Ebola as a Global Health Security Threat: From 'Lessons Learned' to Coordinated Multilateral Containment Efforts. *Journal of Global Infectious Diseases* 6 (4): 164–177.

Kaner, Jolie, and Sarah Schaack. 2016. Understanding Ebola: The 2014 Epidemic. *Globalization and Health* 12: 53.

Lancet. 2004. Statement by the Editors of the Lancet. *Lancet* 363 (9411): 820–821.

———. 2010. Retraction—Ileal-lymphoid-nodular Hyperplasia, Non-specific Colitis, and Pervasive Developmental Disorder in Children. *Lancet* 375 (9713): 445.

Langmuir, Alexander D. 1976. William Farr: Founder of Modern Concepts of Surveillance. *International Journal of Epidemiology* 5 (1): 13–18.

Léonard, Marie des Neiges. 2015. Who Counts in the Census? Racial and Ethnic Categories in Francis. In *The International Handbook of the Demography of Race and Ethnicity*, ed. Rogelio Sáenz, David G. Embrick, and Néstor P. Rodríguez, 536–552. Dordrecht: Springer.

Lloyd-Price, Jason, Anup Mahurkar, Gholamali Rahnavard, Jonathan Crabtree, Orvis Joshua, A. Brantley Hall, Arthur Brady, Heather H. Creasy, Carrie McCracken, Michelle G. Giglio, Daniel McDonald, Eric A. Franzosa, Rob Knight, Owen White, and Curtis Huttenhower. 2017. Strains, Functions and Dynamics in the Expanded Human Microbiome Project. *Nature* 550: 61–66.

Martinez, J. Michael. 2012. Managing Scientific Uncertainty in Medical Decision Making: The Case of the Advisory Committee on Immunization Practices. *Journal of Medicine and Philosophy* 37 (1): 6–27.

Massad, E., F.A.B. Coutinho, M.N. Burattini, and L.F. Lopez. 2004. The Eyam Plague Revisited: Did the Village Isolation Change Transmission from Fleas to Pulmonary? *Medical Hypotheses* 63 (4): 911–915.

Maxmen, Amy. 2010. World's Second-Deadliest Ebola Outbreak Ends in Democratic Republic of the Congo. *Nature* (June 26) [online]. https://www.nature.com/articles/d41586-020-01950-0. Accessed 11 July 2020.

McClain, Charles. 1986. Of Medicine, Race, and American Law: The Bubonic Plague Outbreak of 1900. *Law and Social Inquiry* 13: 447–513.

Molina, Natalia. 2006. *Fit to be Citizens? Public Health and Race in Los Angeles, 1879–1939.* Berkeley: University of California Press.

Murch, Simon H., Andrew Anthony, David H. Casson, Mohsin Malik, Mark Berelowitz, Amar P. Dhillon, Michael A. Thomson, Alan Valentine, Susan E. Davies, and John A. Walker-Smith. 2004. Retraction of an Interpretation. *Lancet* 363 (9411): 750.

Navin, Mark. 2013. Competing Epistemic Spaces: How Social Epistemology Helps Explain and Evaluate Vaccine Denialism. *Social Theory and Practice* 39 (2): 241–264.

Neill, Deborah J. 2012. *Networks in Tropical Medicine: Internationalism, Colonialism, and the Rise of a Medical Specialty, 1890–1930.* Stanford: Stanford University Press.

Northern Ireland Statistics and Research Agency (NISRA). 2019. History of the Census. https://www.nisra.gov.uk/statistics/census/history-census. Accessed 11 July 2020.

Office for National Statistics. 2015. Census History. http://www.ons.gov.uk/ons/guide-method/census/2011/how-our-census-works/about-censuses/census-history/index.html. Accessed 11 July 2020.

Parkinson, Norman. 2013. Thomas Fresh (1803–1861), Inspector of Nuisances, Liverpool's First Public Health Officer. *Journal of Medical Biography* 21 (4): 238–249.

Pica, Natalie, and Florence Bourgeois. 2016. Discontinuation and Nonpublication of Randomized Clinical Trials Conducted in Children. *Pediatrics* 138 (3): e20160223.

Plutynski, Anya. 2012. Ethical Issues in Cancer Screening and Prevention. *Journal of Medicine & Philosophy* 37 (3): 310–323.

Rao, T.S. Sathyanarayana, and Chittaranjan Andrade. 2011. The MMR Vaccine and Autism: Sensation, Refutation, Retraction, and Fraud. *Indian Journal of Psychiatry* 53 (2): 95–96.

Randall, David K. 2019. *Black Death at the Golden Gate: The Race to Save America from the Bubonic Plague.* New York: W.W. Norton & Company.

Rassy, Dunia, and Richard D. Smith. 2013. The Economic Impact of H1N1 on Mexico's Tourist and Pork Sectors. *Health Economics* 22 (87): 824–834.

Reeves, Jay. 2017. Tuskegee Syphilis Study descendants seek settlement money. Associated Press (July 15) [online]. https://globalnews.ca/news/3601333/tuskegee-syphilis-study-suit/. Accessed 11 July 2020.

Reich, Jennifer A. 2016. Of Natural Bodies and Antibodies: Parents' Vaccine Refusal and the Dichotomies of Natural and Artificial. *Social Science & Medicine* 157: 103–110.

Reverby, Susan M. 2009. *Examining Tuskegee: The Infamous Syphilis Study and Its Legacy.* Chapel Hill: University of North Carolina Press.

Roberts, Leslie. 1989. Disease and Death in the New World. *Science* 246: 1245–1247.

Rogers, Katie, and Apoorva Mandavilli. 2020. Trump Administration Signals Formal Withdrawal From W.H.O. *The New York Times* (July 7) [online]. https://www.nytimes.com/2020/07/07/us/politics/coronavirus-trump-who.html. Accessed 11 July 2020.

Saposnik, Gustavo, Donald Redelmeier, Christian C. Ruff, and Philippe N. Tobler. 2016. Cognitive Biases Associated with Medical Decisions: A Systematic Review. *BMC Medical Informatics Decision Making* 16 (1): 138.

Schwartz, Jason L., Arthur L. Caplan, Ruth R. Faden, and Jeremy Sugarman. 2007. Lessons from the Failure of Human Papillomavirus Vaccine State Requirements. *Clinical Pharmacology & Therapeutics* 82 (6): 760–763.

Shen, Ye, Marcy Winget, and Yan Yuan. 2018. The Impact of False Positive Breast Cancer Screening Mammograms on Screening Retention: A Retrospective Population Cohort Study in Alberta, Canada. *Canadian Journal of Public Health* 108 (5–6): e539–e545.

Siena, Kevin P. 1998. Pollution, Promiscuity, and the Pox: English Venerology and the Early Modern Discourse on Social and Sexual Danger. *Journal of the History of Sexuality* 8 (4): 553–574.

Smith, Charles B., Margaret P. Battin, Jay A. Jacobson, Leslie P. Francis, Jeffrey R. Botkin, Emily P. Asplund, Gretchen J. Domek, and Beverley Hawkins. 2004. Are there Characteristics of Infectious Diseases that Raise Special Ethical Issues? *Developing World Bioethics* 4 (1): 1–16.

Stolberg, Sheryl Gay. 2020. Trump Administration Strips C.D.C. of Control of Coronavirus Data. *The New York Times* (July 14) [online]. https://www.nytimes.com/2020/07/14/us/politics/trump-cdc-coronavirus.html?utm_source=nl&utm_brand=wired&utm_mailing=WIR_Science_071520&utm_campaign=aud-dev&utm_medium=email&utm_term=WIR_Science&bxid=5c92c787fc942d0bdf041dba&cndid=14436686&esrc=bounceX&source=EDT_WIR_NEWSLETTER_0_SCIENCE_ZZ. Accessed 15 July 2020.

Tayman, John. 2006. *The Colony*. New York: Scribner.

Thomas, Ashley J., P. Kyle Stanford, and Barbara W. Sarnecka. 2016. No Child Left Alone: Moral Judgments About Parents Affect Estimates of Risk to Children. *Collabra* 2 (1): 20. https://doi.org/10.1525/collabra.33. Accessed 11 July 2020.

Tilley, Helen. 2016. Medicine, Empires, and Ethics in Colonial Africa. *AMA Journal of Ethics* 18 (7): 743–753.

Tognotti, Eugenia. 2013. Lessons from the History of Quarantine, from Plague to Influenza A. *Emerging Infectious Diseases* 19 (2): 254–259.

Tversky, Amos, and Daniel Kahneman. 1973. Availability: A Heuristic for Judging Frequency and Probability. *Cognitive Psychology* 5 (2): 207–232.

UCLA Department of Epidemiology. 2015. Competing Theories of Cholera. http://www.ph.ucla.edu/epi/snow/choleratheories.html. Accessed 11 July 2020.

United States Census Bureau. 2018. Race: About (Jan, 23). https://www.census.gov/topics/population/race/about.html. Accessed 11 July 2020.

United States Census Bureau (Census Bureau). 2015. 1790 Overview. https://www.census.gov/history/www/through_the_decades/overview/1790.html. Accessed 11 July 2020.

———. 2019. Privacy & Confidentiality. https://www.census.gov/history/www/reference/privacy_confidentiality/. Accessed 11 July 2020.

Wakefield, A.J., S.H. Murch, A. Anthony, J. Linnell, D.M. Casson, M. Malik, A.P. Dhillon, M.A. Thomson, P. Harvey, A. Valentine, S.E. Davies, and J.A. Walker-Smith. 1998. Ileal-lymphoid-nodular hyperplasia, Non-specific Colitis, and Pervasive Developmental Disorder in Children. *Lancet* 351 (9103): 637–641. (retracted).

Wong Wai v. Williamson, 103 Fed. Rep. 1 (1900).

World Health Organization (WHO). 2011. Implementation of the International Health Regulations (2005): Report of the Review Committee on the Functioning of the International Health Regulations (2005) in relation to Pandemic (H1N1) 2009. http://apps.who.int/gb/ebwha/pdf_files/WHA64/A64_10-en.pdf?ua=1. Accessed 12 July 2020.

———. 2017a. Plague. http://www.who.int/mediacentre/factsheets/fs267/en/. Accessed 12 July 2020.

———. 2017b. WHO Guidelines on Ethical Issues in Public Health Surveillance. http://apps.who.int/iris/bitstream/handle/10665/255721/9789241512657-eng.pdf?sequence=1. Accessed 12 July 2020.

Chapter 3
Case Identification and Contact Tracing

3.1 Background

Identifying cases of contagious disease and following chains of transmission stemming from them is a traditional mainstay of public health efforts. For example, the first "Healthy People" report of the U.S. Surgeon General explained in 1979—just before awareness of the arrival of HIV/AIDS—that "surveillance—a basic tactic for disease control" requires finding cases of disease or significant exposure, reporting these cases, determining the implications of these reports, and responding appropriately with control measures (DHHS 1979, 9–29). With COVID-19, case identification and contact tracing have been relied upon extensively although they have been less effective where infection has already spread widely.

Such case identification and contact tracing may interrupt chains of infection transmission, protecting both individuals and society more broadly. But it may be seen as intrusive and risky for identified individuals who may be shunned, quarantined, or condemned. At first glance, the strategy of finding people who may spread disease presents a conflict between the overall social good and the interests or rights of the unfortunate individual. This chapter argues, however, that pitting case identification and contact tracing as the individual against society is far too simple. If individual cases are taken seriously not only as potential spreaders of disease but also as people who may be harmed by illness or quarantine, the ethical conflicts posed by their identification can be mitigated in ways that may also enhance the strategy's success.

In this chapter, we attempt to present a balanced case for contact tracing. One disadvantage is that it may be too little too late. Another is that it may inflict significant harm on individuals, as the story of Typhoid Mary illustrates. It also may reflect prejudices and predominant social mores. Case identification has darker sides of prejudice, isolation, and condemnation—that sustain arguments for insisting on individual informed consent before examination and treatment, as well as for

J. G. Francis, L. P. Francis, *Sustaining Surveillance: The Importance of Information for Public Health*, Public Health Ethics Analysis 6, https://doi.org/10.1007/978-3-030-63928-0_3

protection of patient confidentiality. Still, contact tracing may be one of the most important methods for public health to identify and contain disease spread. It is best justified, however, when its benefits are widely known and acted upon, when it is conducted justly, and when its harms are fully addressed.

3.2 Typhoid Mary and Case Identification

Typhoid Mary was a famous example of the success of case identification to discover an individual as a source of infection spread. Her story continues to provide lessons for surveillance today. On the one hand, she demonstrates a clear achievement of case identification. On the other hand, she reveals how treating these cases as mere subjects of infection dehumanizes both sides and may be complicated by racism and opprobrium. Case identification also focuses on the individual rather than on the social circumstances in which she is found. Sustaining surveillance requires respect and equitable and humane responses to those identified as sources of infection. It also requires recognizing and addressing the social circumstances and potentially also injustice in which these cases of infection occur.

According to George Soper (1939), the epidemiologist who identified her, Mary Mallon was a household cook in New York in the early decades of the twentieth century. Investigation of a typhoid outbreak at the summer home of the wealthy Warren family on Oyster Bay, Long Island, suggested Mallon as the cause. Tracing Mallon's work history identified seven other socially prominent households with unexpected typhoid cases while Mallon was employed as their cook. When Mallon resisted examination, she was arrested and detained. Bacterial cultures indicated conclusively that she was a symptom-free carrier of typhoid. Based on these findings, Mallon was locked up as a danger to society. She was released almost three years later, after pledging that she would give up her profession as cook, take other precautions, and report regularly to the Department of Health—commitments that she immediately violated. Mallon was located again five years later, when a typhoid outbreak at a women's hospital was traced to her as the cook. This time, she was confined indefinitely on an isolated island where she remained until her death twenty-three years later. Ultimately, 53 typhoid cases and three deaths were linked to Mallon.

Mallon's lodging during her final long confinement was a pleasant bungalow. She was paid for doing laboratory work and was occasionally allowed to visit nearby New York City, but she was otherwise isolated. Her isolation was painful to her and her treatment remains controversial. Novels have personalized Mallon's own grief in contrast to depictions of her as an unfeeling monster (Keane 2013). Mallon was Irish and Irish immigrants were disfavored at the time as poor and Catholic. Historians have documented the influence of these anti-immigrant and classist sentiments on her pursuit and confinement (Leavitt 1997). By the time Mallon died, New York health officials had identified about 400 other asymptomatic carriers of typhoid, none of whom were reportedly confined in the way Mallon was (Marineli et al. 2013).

Mallon's identification and management were characteristic of U.S. public health activities at the time, although the extent of her confinement was extreme. Addressing public health concerns about typhoid fostered the professional organization of public health activities. The American Public Health Association (APHA), formed in 1872, acquired the *Journal of the Massachusetts Boards of Health* in 1911 and began publishing it as the *American Journal of Public Health* (*AJPH*), a journal still in publication today. The APHA sought to professionalize public health as a means to address the elevated rates of typhoid in the US, rates far higher than in Europe. The APHA lent its support to aggressive anti-typhoid efforts, including education, physician reporting, and carrier identification and containment. The well-publicized discovery of Mallon and the resulting success in stopping her from continuing to infect others lent impetus to these efforts.

Not incidentally, Mallon's identification also played an important scientific role in confirming the germ theory of disease and subsequent scientific developments based on it. By the time her case came to light in the beginning of the twentieth century, the miasma theory of disease—which attributed transmission to the unhealthy odors prevalent in crowded urban areas—was increasingly being supplanted by improved scientific understanding of the role of microorganisms like the typhoid bacillus. In the U.S., state health departments had established scientific laboratories to analyze the presence of infectious diseases, first in Massachusetts where W.T. Sedgwick developed methods for identifying fecal bacteria in water as the cause of typhoid fever during the 1880s (Committee 1988, p. 63). Cities and states established disease registries and reporting requirements, not only for contagious conditions but also for cancer (Committee 1988, p. 66). Over the course of the twentieth century, prevention and treatment improved greatly as well, opening new possibilities for the success of case identification and contact tracing methods. Sulfa drugs that could be used to treat diseases such as pneumonia and gonorrhea came into use in the late 1930s and penicillin in the 1940s. Vaccination against tetanus became routine in the 1940s and the Salk polio vaccine was introduced in the mid-1950s. Gamma globulin therapy as an immune booster for people who had suffered exposures to contagious disease was first used in the 1930s. And these are but a few examples.

With this burgeoning scientific knowledge, public health efforts increasingly focused on controlling the spread of disease from individual to individual. Public health officials directed their efforts to finding individual cases of disease and tracing their contacts so that preventive measures and any necessary treatment could be put into place. Amy Fairchild and colleagues describe how the professionalization of public health was linked to growing knowledge of bacteriology (Fairchild et al. 2007, p. 3).

According to its critics, however, the shift from the miasma theory to the germ theory of disease encouraged public health activities to move in directions that were not always beneficial in improving overall health. The community improvements of the sanitarians gave way to assignment of individual responsibility for disease, often tinged with moralism and racial or cultural critique. U.S. public health largely with-drew from the fields of public sanitation improvements and social reform that had been so successful and moved instead to detecting individual targets as sources of disease and aiming measures at these individuals to prevent disease spread. Education in personal and domestic hygiene loomed increasingly important as a public health strategy. Self-improvement rather than social improvement became the prescribed norm. The historian Nancy Tomes (1997) relates how the change in understanding of tuberculosis from hereditary weakness to chronic contagious dis-ease spurred efforts at household cleanliness and the improvement of an individual's constitution to resist disease. Bodily discharges were viewed as the source of infec-tion and home hygiene as critical to prevention. While Progressive social reformers at the turn of the twentieth century debated the role of substandard housing and poverty in disease causation, Tomes contends (286), they also considered high rates of tuberculosis "as a poor reflection on the cleanliness and temperance of specific ethnic and racial groups," sentiments that, as we discuss later in this chapter, also influenced attitudes towards sexually transmitted diseases. Such sentiments have not entirely disappeared, as revealed by the remarks of an Ohio legislator who attributed racial differences in COVID-19 infection rates to poor hygiene practices. Reportedly, he was fired from his position as an emergency physician because of his overtly racist speculation (Gabriel 2020).

Medical historian Allan M. Brandt adds the changing nature of the American family and the interest in eugenics to this picture. By the turn of the twentieth cen-tury, Americans were having fewer children, divorce had become more common, and growing numbers of women were pursuing careers (1985, p. 7). The sociologist E.A. Ross had proposed the theory of "race suicide" in 1901: that as Anglo-Saxon families were gradually having fewer and fewer children, they were becoming over-whelmed in numbers by children of color and children of immigrants. Brandt relates how the resulting anxiety spurred efforts both to reduce the fertility rates of those judged to be less desirable through eugenic measures and to increase the fecundity of those judged reproduction-worthy through health promotion measures including venereal disease control. Growing medical knowledge about the transmission of sexually transmitted infections (STIs) and their role in causing infertility lent urgency to efforts to identify the infected and either treat them or prevent them from engaging in behaviors believed to risk disease spread. These medical efforts inter-twined with the moral attitudes and social circumstances of the times in shaping approaches to disease identification and control.

As a result of this shift in responsibility for disease prevention from the social to the individual, public health in the U.S. did not return to playing a significant role in promoting workplace safety or environmental protection until late in the twentieth century (Fairchild et al. 2010). COVID-19 outbreaks in meat packing plants laid bare continuing deficiencies in workplace safety. Nor did public health address the social determinants of health—that is, the social conditions that affect the health of popula-tions—in ways that it might otherwise have done during much of the twentieth cen-tury. Instead, U.S. public health located disease causation and remediation, along with the potential for responsibility and blame, squarely in individuals. This individualist paradigm for public health can be viewed as part of a more general U.S. emphasis on individual responsibility for health and on health care for individuals rather than public health measures. Typhoid Mary was emblematic of this individualism: a case of disease to be found, condemned for the harm she had caused, and isolated.

3.3 Contact Tracing

Contact tracing starts with an "index case" like Typhoid Mary, a person with a contagious disease who has become known to public health authorities. (An index case may not be a "primary case," the first person through whom a disease is introduced into a population; and a primary case may not be such an index case unless it becomes known to public health.) The person is then interviewed, asked about contacts to whom the disease might have been transmitted, and treated or isolated if necessary. Then, public health authorities follow up by interviewing and testing the named contacts, treating or isolating them if necessary, and asking them about any further contacts. The chains may break, and thus the method lose efficacy, if the index case cannot be found or interviewed or does not name all relevant contacts, or if the contacts in turn cannot be found or do not give information.

Even when contact tracing works reasonably well, it has advantages and disadvantages as a method for reducing disease spread. It does find cases that may not otherwise have been recognized and allows for their education, isolation, or ideally treatment. Yet it may be expensive and ineffective in altering spread, depending on a disease's mode of transmission. Its efficacy also depends on the stage of disease spread within a population; it is most effective in the early stages of an outbreak or when other protective measures have reduced the numbers of new cases to a manageable level (Moore et al. 2020). It is intrusive and may generate resentment and secrecy, ultimately undermining its very aims if people refuse to cooperate.

The method of case identification and contact tracing is simplest and least expensive if specific types of easily recognized encounters are required for disease transmission. Sexual intercourse is an obvious example: people are likely to know that it has happened and the identity of the person with whom it occurred. Contagious diseases that are transmitted by airborne droplets are not as susceptible to the methodology, however. For example, someone who rode on a bus may not remember sneezing during the ride and surely is unlikely to know who else was on the bus. Nor can contact tracing be readily used for diseases that are transmitted by fomites, surfaces such as countertops or doorknobs that can harbor infectious agents. The length of time an infectious agent may survive on a fomite, moreover, may vary by factors such as temperature and humidity (Boone and Gerba 2007).

Coronavirus, different types of which can cause infections from colds to SARS to COVID-19, is an example of an infectious agent that can be transmitted by both airborne droplets and fomites. Successful tracing would need to follow the movements of the index case and find anyone who has been in an area or had contact with any surface for some period of time after it had been touched by the index case. Technologies such as location tracking have been called on to help fill in these gaps with COVID-19, but their efficacy and acceptability remain in question, as we discuss later in this chapter. It is thus not surprising that infectious diseases that are transmitted by sexual contact or direct contact with bodily fluids have drawn the bulk of public health interest in contact tracing as a surveillance method. But mode of transmission is not the only reason for this attention: moral condemnation of those transmitting diseases through sex has also played a major role in how these diseases have been judged and managed by public health.

Even when it may be effective in following paths of disease transmission so that new cases can be identified and treated (or, if treatment is not possible, isolated), surveillance through contact tracing is time-intensive and intrusive. The process takes resources even if the contacts of the index case are known, identified, relatively easy to locate, and readily treatable. Moreover, when a disease has a long incubation period, contacts may need to be followed for an extended time to determine whether they have become infected. The incubation period for influenza is comparatively short—an average of two days—but the incubation period for measles averages 10–12 days and the period for Ebola may be as long as 21 days. Although our knowledge is still evolving, for COVID-19 it may take up to 14 days for individuals to develop symptoms, and people may be at their most contagious before any symptoms appear. Estimates are that as much as 40% of transmission may come from people who have not yet developed symptoms or who may never become symptomatic (Moore et al. 2020). Effective testing programs may therefore need to test both symptomatic and asymptomatic people. Incubation periods for other diseases may be far longer: the incubation period for mononucleosis can be a month or two and for tuberculosis as long as six months. Methods of testing may be embarrassing or intrusive, ranging from cheek swabs to swabs high up in the nasal cavity, pinpricks to blood draws, and excrement collection to biopsies. Applied mathematics can be used to model the comparative cost-effectiveness of contact tracing for diseases with these and other varying characteristics (Armbruster and Brandeau 2007). When diseases are not particularly serious, however, contact tracing may not be worth its costs.

Contact-tracing can be targeted in ways that are increasingly cost-effective, such as by concentrating on high risk exposure situations like nursing homes or prisons (Moore et al. 2020). Targeted tracing will still involve intruding on at least some individuals identified as possible disease sources, however. Contacts who are highly socially networked and who thus may have infected many others will be particularly valuable to find from the perspective of cost-effective reduction in disease spread.

In addition, when a disease has become widely established in a population, contact tracing likely is not the best strategy for addressing it (Moore et al. 2020). If a very high percentage of people in a population are already infected, contact tracing may simply put effort into finding people who have already been infected from other sources. Approaching everyone in the population directly with education and offers of available treatment could potentially be far more effective in reducing infection rates than following chains from index cases. So might other disease mitigation strategies such as the mask wearing recommended for COVID-19. Such more universal approaches also have the advantage of detaching the presence of infection from its source and thus from any suggestion that blame might be attributed to individuals suspected of transmission. Universal education campaigns, however, have sometimes met with resistance precisely because of this detachment. The history of case identification and contact tracing has—not surprisingly—been linked to diseases drawing moral opprobrium, especially those that are transmitted through sex, or associated with members of disfavored groups such as Typhoid Mary.

Finally, when people do not trust the government, and identify public health with the government, contact tracing may be very difficult. Some of the surge of the COVID-19 pandemic in the United States coincided with the murder of George Floyd by police officers and the Black Lives Matter protests that followed. Initial reports suggested that tracers were successful in getting names of contacts from fewer than half of those identified as infected with COVID-19 in New York and other major U.S. cities (Otterman 2020).

3.4   Progressivism, Moral Purity, and Sexually Transmitted Infections

Sexually transmitted infections (STIs) such as syphilis and gonorrhea were at the heart of case identification and contact tracing as the methods developed. This history is important not to forget even as contact tracing is deployed for contagious diseases that do not carry this particular stigma, such as COVID-19. Not surprisingly, the methods were tinged with the moral judgments associated with these diseases—and may still be tinged with finger-pointing or worse in some societies. Case identification and contact tracing were also associated with the Progressive movement in the United States—a movement that, as its name suggests, was attracted to social and scientific progress but also with eugenics and the moral rectitude of many of its proponents. In the context of this history, case identification and contact tracing were primarily about protecting society from diseases identified with moral failure, rather than protecting its victims from social ills. As mores change, however, contact tracing may become more—or less—acceptable for different diseases.

Known for centuries, syphilis may have arisen in the Americas, as contemporary methods of skeletal analysis now suggest, and been carried back to Europe by explorers from Spain—an illustration from an earlier time period of the role of global travel in the transmission of disease (Rothschild 2005). However, other explanations also place syphilis in Europe before exploration of the Americas. Soldiers throughout history have been a primary source of syphilis transmission. After Charles VIII of France invaded Naples in 1494, returning soldiers brought a virulent form of the disease back with them (Frith 2012). Metals such as mercury were used in hopes of treatment but with highly toxic side effects.

Medical knowledge about STIs changed radically as the nineteenth century moved into the 20th. Gonorrhea and syphilis were identified as different diseases with specified bacterial causes. Gonorrhea's longer-term effects such as pelvic inflammation and scarring in women, infertility in men, or eye infections and blindness came to be recognized. The early stage of syphilis infection was linked to sequelae such as the nerve or joint damage of the disease in its tertiary stage. The first effective treatment for syphilis—an arsenic-based form of chemotherapy known as salvarsan, followed several years later by the less-toxic but nonetheless grueling neo-salvarsan—was introduced about 1910; it required a long course of injections and had significant side effects.

The Crimean War in the 1850s, the U.S. Civil War in the 1860s, and then World War I, had each brought new waves of syphilis transmission to domestic populations. Faced with long periods of time away from families or friends and deployments that were often tedious and more often unpleasant or frightening, soldiers unsurprisingly sought out prostitutes or sex. Countries with soldiers stationed overseas took opposing approaches to the problem of disease transmission through prostitution. Some—Britain is an example—pursued primarily a regulatory strategy, seeking to ensure that prostitutes remained "clean" of disease. Others—the U.S. is a notable example—pursued a strategy of condemning prostitution and attempting (without apparent success) to discourage its soldiers from seeking commercial sex. According to Frith (2012), STIs were second only to influenza for soldiers' lost days for duty during World War I, accounting for nearly seven million days in which soldiers were unavailable. Clearly the military had incentives to make efforts to reduce the impact of these diseases, and civilian populations did as well.

The first legislation regulating STIs was the UK Contagious Diseases Acts of 1864 and 1866. Impelled by high rates of STIs among soldiers returning from the Crimean War, the Acts required compulsory registration of prostitutes and their police supervision, regular examination, and hospital detention of prostitutes found to be infected (Adler 1980). The strategy of the Acts was to permit prostitution but subject those engaging in it to compulsory management. This approach drew sharp criticism from early feminists and advocates of moral purity in the late Victorian era. Although the Acts were repealed in Britain in 1886 as part of efforts to abolish prostitution, historian Philippa Levine (1996) describes how similar regulatory strategies remained in play in British colonies such as India, where imperialists urged their continuation for the protection of British soldiers and British military power. Britain deployed a force of some 60,000 soldiers in the direct rule of India, soldiers who were thought to require some sexual services from prostitutes. Disease control measures were imposed on prostitutes and not on the soldiers who frequented them, however, to the chagrin of incipient nationalist movements. Regulatory strategies were also pursued in France and in Germany, with prostitution limited to specified areas of cities, registration and examination of prostitutes, and arrests and fines for those plying the trade clandestinely (Brandt 1985, 35). By contrast, with minimal exceptions of initiatives led by physicians who had been educated in Europe, most U.S. jurisdictions prohibited prostitution rather than attempting regulation. Nevada was an exception, reflecting the openness of the west and Gold Rush days, but even Nevada regulated prostitution and confined it primarily to smaller towns and rural areas (Symanski 1974). Becoming STI-infected was viewed as punishment for promiscuity. Moralists argued that regulation would encourage sin rather than protecting against its resulting harms. The U.S. Mann Act, criminalizing transportation of women across state lines for "immoral purposes," was enacted in 1910, just before World War I. Regulation of STI transmission was thus a policy stage on which conflicting views about colonialism, gender, and moral purity played out.

The world wars of the twentieth century likewise brought clashing views about STI control into play. Here, moralism in the U.S. was especially strong. During World War I, while some nations tried to protect their soldiers with condoms or to encourage supposedly safer forms of prostitution, the U.S. military instituted chastity campaigns. Posters graphically depicted the ravages of syphilis and gonorrhea and their ability to hide in the bodies of enticing but unclean women. The sex education film "Fit to Fight," shown to recruits, urged soldiers to resist sex in support of the war effort and portrayed prostitutes as enemy agents undermining military resolve (Brandt 1985, 67). Anti-alcohol campaigns became part of the mix, too, as prostitution was associated with bars and saloons; prohibition itself was the law of the land in the U.S. from 1920 to 1933. Training camps for soldiers featured entertainment, organized athletics, and chaperoned visits from wives and girlfriends. As medical historian Brandt describes (1985), however, medical testing of soldiers revealed that over 10% of newly enlisted soldiers were already infected with STIs and that rates in some camps were as high as 25%. The shock of these statistics fueled social hygienists' pursuit of U.S. national efforts against venereal disease and alcohol.

Efforts against STIs and related "immorality" only intensified when U.S. soldiers nonetheless quite predictably returned from the first world war with infections. Because under the U.S. federal system responsibility for public health lies largely at the state level, states were in the forefront of these efforts. Some states responded with coercive measures to prevent disease spread.

These responses were intertwined with the politics of the day, including Progressivism and feminism. The Progressive era in U.S. politics is generally considered to have extended from the 1890s through the end of the first world war. Spurred by the unfettered growth of capitalism, shifts from rural life to urban industrialization, and increasing social inequality, Progressivism opposed corruption and favored the introduction of social science methods into government. Progressives pushed for economic and social reform, including measures such as worker's compensation and social security.

But there were morally grimmer sides to Progressivism: racism, eugenics, and opposition to immigration. Many Progressives advocated social Darwinism, promoting birth control and mandatory sterilization of those believed to be unfit to promote healthy births for a healthy population. Economist Thomas Leonard describes these "illiberal" tendencies of the Progressive era as paradoxically both rooted in social science and failing to apply the insights of social science to understand the potential biases of Progressivism itself. He writes, "Progressive Era reform at once uplifted and restrained, and did both in the name of progress. In practice, only white men of Anglo-Saxon background escaped the charge of hereditary inferiority, and even members of this privileged group were condemned as inferiors when they, as with *The Jukes* and other 'white trash' families studied by eugenicists, were judged deficient in intellect and morals" (Leonard 2016, xiii). Sex outside of marriage and sexually transmitted infections were considered marks of unfitness. Moreover, preventing people from breeding was cheaper and simpler than supporting them and their offspring.

Progressivism was also associated with paternalism in the pursuit of the betterment of individuals. Although the typical characterization of the Progressive era is that it featured a shift from rural to urban politics, sociologist Kristin Luker argues that the movement can be better understood as the shift from "familial patriarchy" to "social patriarchy" (1998, p. 601). Groups such as the American Social Hygiene Association and the Woman's Christian Temperance Union, along with many women's clubs across the country, advocated a single standard of sexual abstinence until marriage. Their success in defending women's equality was modest, however; and, in Luker's view, this purity movement was largely amalgamated in the service of medicalized social hygiene directed primarily towards the regulation of female sexual conduct. Laws against prostitution were enacted or strengthened and a number of new women's prisons or reformatories were constructed. By 1918, according to Luker, 32 states had compulsory quarantine laws for venereal disease and federal government statistics indicated that 30,000 women were taken into custody during the 27 months of U.S. involvement in World War I. In her doctoral dissertation, sociologist Nicole Perry (2015) describes how Kansas quarantined over 5000 women, primarily low-income and working class and suspected of having STIs, in a prison farm during the 1920s and 1930s.

Beginning in the 1890s with Connecticut, many U.S. states passed laws regulating who could marry. These laws were initially eugenic, prohibiting the "feeble minded," "imbeciles," epileptics, and others regarded as of unsound mind from marrying. Not long after, they were joined by laws restricting marriage when people tested positive for venereal disease. The statutes used "venereal disease" to refer to these infections, with the connotations of illicit love that this terminology suggested. Washington's comprehensive statute enacted in 1909, for example, prohibited marriage by anyone who was a "common drunkard, habitual criminal, epileptic, imbecile, feebleminded person, idiot or insane person, or person who has theretofore been afflicted with hereditary insanity, or who is afflicted with pulmonary tuberculosis in its advanced stages, or any contagious venereal disease…" (Lindsay 1998). By the outset of World War II, twenty US states required STI testing for issuance of a marriage license and nineteen required prenatal testing of pregnant women. These statutes varied from primarily health-promoting goals to goals that were more frankly punitive. In Virginia only, those who tested positive were permitted to marry without a waiver from the state health department, so long as the partner was fully informed and agreed to the marriage and both agreed to treatment recommended by the health department. The rationale for the Virginia approach was that it would encourage more people to come forward for testing. The state was concerned that its residents had been avoiding the testing requirement by marrying out of state or by continuing in what the state regarded as an illicit relationship (DePorte 1941). Other states flatly prohibited marriage for those identified as infected unless a waiver was granted by public health authorities. Although rates of positive tests were lower than expected, some observers hypothesized that the explanation was that people who knew or suspected that they were infected avoided the testing (Brandt 1985, 149). States also enacted statutes mandating prenatal testing of pregnant women; these laws revealed higher rates of infection than the premarital testing laws and resulted in significantly lower rates of infants born with congenital syphilis.

This reformist but repressive zeal targeted STIs, especially syphilis, for case identification and contact tracing. Case identification and contact tracing were widely enacted into state law and became the central method for STI control in the U.S. (Brandt 1985; Gostin and Hodge 1998). The moral opprobrium directed towards illicit sex played an important part in shaping the structures for reporting and tracing individuals judged responsible for STI transmission, especially in the U.S. but elsewhere as well. Condemnation of prostitution or sex outside of marriage, opposition to birth control, and public health efforts to reduce burdens of disease intertwined as the structure for implementing these methods developed. Social class played a role in enforcement; according to Fairchild et al. (2007, p. 9), physicians exercised discretion in favor of their more privileged patients in making decisions about when to report cases to public health officials. Racism was part of the mix, too, as blacks and immigrants were considered more likely to be promiscuous and thus infected.

In the U.S., Thomas Parran was the leading advocate of case identification and contact tracing for STI control. His views were complex and can be more readily aligned with public health goals of disease prevention than with the social moralism of many other proponents of case identification. However, Parran was at best an ambiguous figure for the future of public health. As Chapter 2 describes, Parran is now thought to have been associated with the early stages of the Tuskegee syphilis study that remains a symbol among U.S. Blacks for mistrust of public health interventions. Parran began his career in rural sanitation and then became chief of the U.S. Public Health Service's Division of Venereal Disease in 1926, where "he worked to sway public sentiment away from moral condemnation of venereal diseases and toward consideration of syphilis as a medical condition and threat to public health" (Snyder 1995, p. 630). A challenge for Parran was breaking the veil of silence that had reemerged around venereal disease during the 1920s (Brandt 1985, 122). Federal funding for disease control had fallen off and opposition to treatment had grown, fueled by opposition to the supposedly loose morals of the Roaring '20s and the belief—shared by Parran—that ready availability of prophylaxis would encourage unsafe behavior. As the depression took further toll on federal funding, Parran left the federal government in 1930 to become New York State health commissioner. There, he confronted continued repression of discussions about sex; in 1934, he was not allowed by CBS radio to deliver an address over the radio that mentioned syphilis or gonorrhea (Brandt 1985, 122). He also confronted continued opposition to treatment as encouraging immorality and the stark reality that few people during the depression could afford the costs of care.

Later, as U.S. Surgeon General from 1936 to 1948, Parran was in an especially strong position to campaign against syphilis. Parran "brought a scientific, bureaucratic approach to the venereal problem [a]s a career public health officer" (Brandt 1985, p. 138). Parran's efforts made remarkable inroads into silence about STI infections and especially their impact on women; an article he coauthored in the respected and widely read women's magazine *Ladies' Home Journal* brought information about STIs into highly traditional circles. Importantly, Parran's plan was multifaceted, including testing, treatment, and public education. He urged the establishment of free and confidential testing centers in high risk areas, to be followed promptly by treatment. Parran argued as well that there should be public funding for treating all infected persons.

In 1938, the U.S. Congress enacted the National Venereal Disease Control Act, acknowledging the national scope of the problem of STIs and the need for the federal government to play a significant role in addressing it. The Act included funding for research and for grants to state public health departments to use in developing plans for disease control activities. Funding for treatment was limited, however; one difficulty was avoiding the charge of physicians and the American Medical Association that public funding would be a start on the road to "socialized" medicine. Nonetheless, Parran's efforts achieved notable success: Tramont and Boyajian (2010) cite data that in 1942, according to census figures, 580,000 people were newly infected with syphilis; but after Parran's campaign, the rate of new infections had fallen to 120,000 cases a mere 10 years later. The success is even more impressive given that the only treatment available at the time, neo-salvarsan, was expensive, time-consuming and risky; penicillin became available for treatment of syphilis only in 1942.

These methods were joined in many states by other interventions, such as the duty to report those believed to be infected and the duty to warn their sexual partners. Because in the U.S. responsibility for the public welfare lies primarily with the states, these reporting requirements are typically established by state statutes and implementing regulations. (The federal government might use its power to regulate interstate commerce to regulate disease threats but has not done so; control of the country's external borders does lie with the federal government and it has imposed restrictions on entry for those with specified infectious conditions.) Statutes may list reportable diseases and give public health agencies the authority to identify reportable conditions through administrative rulemaking. These structures will specify what individuals or institutions are required to report, which conditions are reportable, and where reports should be made. States may use these case reports to investigate outbreaks and to follow up contacts of individuals who may be contagious. States today also publish reports of disease trends and, under agreement with the U.S. CDC, share information regarding disease outbreaks and trends (CDC 2019).

The city of Chicago was particularly enthusiastic about implementing syphilis control measures and became a flagship city for the national effort. As Poirier (1995) describes in her comprehensive literary history of the events, in the late 1930s

Chicago instituted a massive campaign to test and treat syphilis. The experience of the Chicago Syphilis Control Program illustrates the methods of case identification and contact tracing at their best and worst.

On the one hand, aided by publicity from the *Chicago Tribune* and other media outlets, the Program achieved high visibility and brought widespread attention to syphilis. Initial letters mailed to one million residents offered free testing that was to be voluntary, conducted by the family's own physician, and with the promise of complete confidentiality about results. To avoid stigma, everyone was invited to be tested; organizations such as the *Tribune* encouraged testing by their employees and meetings of Lions clubs and the American Legion also offered testing. Dr. Ben Reitman, an unconventional physician who periodically rode the railroads home-lessly himself and was the lover of the leftist Emma Goldman, provided treatment and prophylaxis for prostitutes, the homeless, and the poor without question and usually with respect for them and their life circumstances. Treatment was publicly funded for many and these people received medical care that would otherwise have been unavailable to them. Treatment rates were especially high among African Americans in Chicago, who with the public funding had unprecedented levels of access to care. Rates of infection fell in the city; Chicago had the lowest rate of syphilis among those entering the military in WWII (Brandt 1985, 152).

But the Chicago Syphilis Control Program also had more troubling aspects. The tests for syphilis in use at the time had high false positive rates—possibly as high as 25% (Brandt 1985, 152). People who tested positive falsely may have received unnecessary but toxic treatment. A series of *Chicago Tribune* stories revealed physi-cians who set up fraudulent practices supposedly to identify and treat the infected. While testing was voluntary for most Chicagoans, it often was not voluntary for those accused of crime. The Chicago Women's Court ordered testing of anyone charged with prostitution and hospitalization of those with positive test results. Other courts required testing and imposed fines sufficient to keep supposed offend-ers in treatment. Far from the goal of universal testing that would dilute stigma, Poirier contends, these aspects of the Program were applied in discriminatory ways, for example targeting the women selling sex rather than the men buying it as immoral offenders (1995, 85). Free testing was offered, but people were encouraged to visit their private physicians, both to save costs and to avoid opposition from the Chicago Medical Society that free testing would impact their members economi-cally. Cooper (2001) contends that testing was primarily conducted on those access-ing public sources of health care, as people who had resources could pay for private testing or avoid testing altogether. Although reporting rates from private physicians did increase, they remained low and the result was inequitable understanding of the scope of the epidemic and needed responses to it. African Americans and the poor more generally were disproportionately represented in the statistics, thus creating the potential for stigmatization of these groups. Privacy was another concern; employers were encouraged to test their employees and labor advocates rightly feared that people testing positive would be excluded from health insurance plans until they had been treated successfully (as people with positive tests were at the *Tribune* itself), would be charged higher rates for health insurance, and would be denied life insurance. Schools and universities required testing for entrance, with some such as Northwestern refusing admission to those testing positive (Poirier 1995, 131).

Penicillin became available as an effective and comparatively safe treatment for syphilis beginning in 1942. Risks of a positive test could thus be mitigated—and benefits maximized—as people could delay testing until after undergoing treatment or could receive curative treatment after a positive test. Case identification and contact tracing are least burdensome to individuals undergoing them when treatment such as this is readily available, as it became for both syphilis and gonorrhea. In such circumstances, there are direct benefits to the person receiving a positive test result: treatment and cure. Such situations are a double win, both for the person who receives treatment and for others to whom he or she might have passed on disease. At least, they are such a double win if adverse consequences do not accompany knowledge that the treatment has occurred. In contexts in which knowledge of STI treatment brings damaging rejection—divorce, disownment, expulsion from school, excommunication from a religious community, or job loss, for example—patient confidentiality becomes an issue. Patients may reasonably seek to avoid treatment if the resulting harms to them are significant. Others affected either by the non-treatment or by the conduct that resulted in the STI—such as a spouse who does not suspect a partner's unfaithfulness—might reasonably reply that knowledge of the STI and whether treatment has occurred is of critical importance to their health or to the conduct of their lives. Finally, treatment shortages, as occurred with Penicillin G Benzathine, the recommended method for treating syphilis, may also occur and depress access to treatment (Nurse-Findlay et al. 2017).

The period after World War II was in some respects a golden age for antimicrobial treatment, not only of sexually transmitted infections but also of many other infectious diseases. New classes of antibiotics were becoming available and drug resistance was minimal. Case finding opened possibilities for treatment, benefiting individuals as disease victims and remediating them as potential vectors. Conflicts between protecting individuals from harm and reducing disease spread appeared to have largely ended, and an era begun in which the interests of individuals and the public health were fully aligned.

The appearance of HIV/AIDS in the late 1970s, however, brought sobering challenges to this rosy optimism. AIDS also sharpened the ethical conflicts raised by case reporting and contact tracing. On the one hand, the moralism of the Progressive era was long in the past. Supported by the emerging field of bioethics, patient advocates called for respect for the individual autonomy of patients and the recognition of social responsibilities in fighting disease. For quite some time, however, AIDS was met with fear, condemnation of those affected, and the deadly consequences of unchecked disease spread.

3.5   HIV/AIDS: Disease Control and Confidentiality

Retained tissue samples now suggest that sporadic cases of HIV or a closely related infection may have appeared in humans as early as the 1920s but died out without significant spread. How the current form of pandemic HIV infection emerged remains contested; most likely, the primary form of HIV infection emerged in Kinshasa in the Democratic Republic of the Congo from consumption of infected primates. What became the present-day pandemic was initially identified as a mysterious immune deficiency associated with unusual infections such as thrush and pneumocystis pneumonia and rare cancers such as Kaposi's sarcoma. Because many of the initially identified cases in the U.S. occurred in gay men or Haitians, the disease was associated with these groups and further stigmatization of them.

People with AIDS ("acquired immune deficiency syndrome") were subject to devastating infections because of catastrophic falls in levels of the T cells critical to fighting infections. The disease initially killed primarily young, previously healthy men who became infected and by 1994 had become the leading cause of death among young adults in the U.S. The cause of the T cell destruction was identified by 1983 as the HIV virus, dispelling the mystery of causation. Initial tests for HIV became available in 1985; these had high false positive rates and tests with greater specificity were developed subsequently (Alexander 2016). The discovery that HIV is a retrovirus also led to recognition of its transmission only through direct exposure to bodily fluids such as blood or semen. Fears of casual transmission were not easily allayed in the general population, however, and people with HIV were initially barred from schools, employment, athletic competitions, and many other venues.

Worldwide, AIDS spread rapidly through infected blood supplies, sex workers, and maternal-child transmission (Mann 1987). In some areas, the disease virtually eliminated an entire generation of young adults. By 1988, the WHO estimated that between 5 and 10 million people were infected and that the virus would spread to every country (Samuels et al. 1988). By far the greatest burden of disease was in impoverished areas of sub-Saharan Africa and South and Southeast Asia. Even in resource-rich countries such as the U.S., the greatest burdens of disease were among poor and minority populations. Writing in the *American Journal of Public Health* special issue on AIDS in 2002, Richard Parker characterized AIDS as a disease of structural injustice. Yet AIDS was also a disease that struck people at all economic levels and that became a locus for patient activism and gay activism in more prosperous areas.

3.5.1 HIV Disease Control

Not surprisingly, HIV/AIDS brought intense ethical challenges to STI disease control strategies based on case identification and contact tracing. AIDS advocates contended that testing performed without voluntary and fully informed consent violated people's rights to decide what should be done with their bodies. Required case reporting or other disclosures (for example, to sexual partners) were seen as serious violations of confidentiality, with the risk of deterring people from getting tested. Many of the early AIDS patients in the U.S. were in the arts and highly educated and advocacy on behalf of AIDS patients was forceful and effective. AIDS also came on the scene in the era of gay civil rights activism; unsurprisingly, groups such as the Gay Men's Health Crisis formed almost immediately after initial publicity about identification of the illness. AIDS advocates pushed back against anti-gay actions such as closures of the bathhouses in San Francisco, arguing that these venues could become sources not of disease transmission but of education and community support. Advocates also urged access to health care and the right to try therapies that had not yet been approved by the Food and Drug Administration. Activists also argued strongly for informed consent before testing and protection of the confidentiality of test results. In these arguments, they were joined by defenders of autonomy in the emerging field of bioethics which emphasized informed consent and patient confidentiality.

In the initial decades of the pandemic, AIDS patients in the U.S. faced realistic concerns that reporting would subject them to personal condemnation from family and friends, loss of employment, loss of insurance, quarantine, or even criminal prosecution. (Fallone 1988). Before modes of transmission were understood, blood transfusions were a common source of HIV infection; hemophiliacs who were treated with clotting factor made from multiple blood donations were at particularly high risk. Ryan White, one of the first young hemophiliacs diagnosed with HIV (in 1984), was barred from attending school in the town in Indiana where he lived. His fight to return to school gained national attention; even after the Indiana state health commissioner made clear that HIV could not be transmitted by casual contact, some parents pulled their children out of the school when Ryan returned. The federal statute providing funding for HIV treatment was named after Ryan White and enacted shortly after his death in 1990.

HIV infection also played a central role in changes in the U.S. insurance market that froze out people with pre-existing conditions. Patients with HIV were expensive, particularly early in the epidemic, as they had high rates of hospitalization, although they turned out in the longer run to be less expensive than initially predicted (Scitovsky 1988). Insurers increasingly denied coverage to people with HIV or excluded HIV-related illnesses from the coverage they did provide. By 1989, all states but California had laws permitting insurers to test people for HIV as a condition of health insurance (Lambert 1989). Insurers backed away from community rating—the practice of charging premiums based on the risk of average individuals in the community—to experience rating, calculating premiums based on an individual or a group's prior utilization history. Surveillance information about disease prevalence in various populations enabled these actuarial calculations. People with histories of expensive treatment for HIV/AIDS were priced out of the insurance market as a result. So were groups: some insurers denied coverage to groups such as florists that were believed to have high percentages of gay workers at potential risk for HIV. Larger employers moved increasingly towards maintaining the reserves needed to meet their own costs of insurance, thus avoiding state law coverage mandates (Padgug et al. 1993). The U.S. Supreme Court let stand a lower court decision permitting an employer to reduce lifetime coverage for AIDS-related illnesses from $1 million to $5000 when faced with covering an employee with a positive diagnosis (*McGann* 1991). When people with AIDS were no longer able to work, they lost health insurance from their employers. Burdens of providing care fell especially hard on public hospitals and community-funded clinics. Only with the Affordable Care Act (ACA) has the U.S. moved back in the direction of insurance that cannot deny people coverage or set premiums because of conditions they already have, although as late as 2020 these protections remained under legal challenge.

Many jurisdictions adopted punitive stances in fear of what initially was a fatal disease transmitted in ways unknown. As the virus was identified and transmission understood, support for criminalization of HIV transmission was fueled by moral judgments that people with HIV were responsible for their illnesses because of their intravenous drug use or sexual behavior. These statutes typically made it a crime for people who knew they were HIV positive to engage in behavior thought to create risks of exposure to others. These statutes were not based on evidence that they would increase the likelihood that people would learn their HIV status and avoid infecting others. To the extent that they only subjected people who knew their HIV status to punishment, the incentives they created were to avoid knowledge and thus likely were counterproductive. Nor were they based in evidence that they would reduce either rates of HIV or rates of exposure-prone conduct such as needle sharing (Yang and Underhill 2018). But these punitive statutes remain widespread. Indeed, when California modified its law in 2017 to reduce punishment for HIV transmission to a misdemeanor, and then only when the transmission was intentional and to a sexual partner who was unaware of the offender's HIV status, it went against the clear trend of 34 states that still criminalized HIV transmission. Moreover, enforcement of these statutes in the U.S. has been inconsistent and has had a disparate impact on Blacks.

Outside of the U.S., some jurisdictions resorted to quarantine to stem the tide of the HIV pandemic. Cuba began a widespread program of mandatory screening, first among people who had been outside of the country and thus might have been exposed to the disease and then among all hospitalized patients, pregnant women, and people diagnosed with STIs (Bayer and Healton 1989). Observers reported that informed consent was not required, at least initially. People identified as HIV positive were quarantined in sanatoria located away from cities. Their care was paid for, they continued to be paid salaries or a stipend even though they were unable to work, and they were given passes for monitored visits with families and friends. Cuba justified the program as part of its overall public health approach to health care. Cuba did continue to have very low rates of HIV positivity, with the principal risk factor for infection being sexual contacts while abroad. Cuba also had high rates of treatment of those who became infected. Critics pointed out the risk of confinement of people with false positive tests, the great expense of the program, and the serious deprivation of individual rights that it involved (Pérez-Stable 1991). Others argued that the program should be considered more sympathetically, both because of the benefits it achieved in limiting infection rates and because it was conceived primarily as an initial emergency response. By 1990, as treatment became available, Cuba converted to a voluntary educational and day treatment program while allowing residents of sanitoria to continue to live there if they so wished (Anderson 2009). Moreover, observers also reported that Cuba required informed consent to testing and contact tracing, although it placed far more pressure on people to reveal their contacts than might have occurred elsewhere (Anderson 2009).

As international public health responses to AIDS developed, policy makers also urged incorporation of international human rights norms into HIV policy. AIDS policy stimulated the introduction of human rights into public health ethics. Jonathan Mann, initially as director of the WHO Global Program on AIDS, developed a systematic human rights approach to the disease (Fee and Parry 2008). Mann argued for the rights of those who had become infected, most importantly rights to non-discrimination, access to health care, and protection from coercive interventions such as quarantine. Mann believed that public health goals in disease prevention could be brought into harmony with protections of political and social rights. He contended that HIV flourished where inequality prevailed and that the health of the majority could only be protected if the rights of people at risk were respected. After a change in leadership at the WHO, Mann resigned; several years thereafter, he perished in the crash of a SwissAir plane. But Mann left a legacy of the importance of human rights for HIV policy and public health policy more generally (Gruskin 2002, 2002a).

Thus, from the beginning AIDS was associated with stigmatization of groups such as gay men, intravenous drug users, or Haitians—the last because some of the earlier identified cases were in Haitians or those who had visited Haiti (Castro and Farmer 2005; Parker and Aggleton 2003). According to one report, even the contact tracers themselves were stigmatized from mistaken beliefs that they too might have become infected (Kampf 2008). These beginnings shaped HIV advocacy towards the bioethics paradigm of individual informed consent in the U.S., the paradigm we discuss further in Chapter 7. Commentators urged careful counseling and informed consent procedures prior to any HIV testing or research (Gillon 1987). Providers and others interested in HIV prevention also voiced the realistic fear that public awareness of required reporting would deter people from getting tested. As an alternative, they proposed establishment of anonymous testing venues (Kegeles et al. 1990).

3.5.2   Confidentiality and Reporting Test Results

Confidentiality of their medical information was paramount among the protections urged for patients with HIV. As the epidemic grew, public health law scholar Larry Gostin (1986, 227) argued for widespread voluntary testing and complete protection of the confidentiality of test results. British bioethicist and primary care physician Raanan Gillon (1987) argued that except in extraordinary circumstances HIV test results should not be shared by specialists with the patient's primary care physician without the patient's agreement. These "extraordinary" circumstances were the possibility that a patient with no regard for others would put innocents at risk of infection. Even in such circumstances, Gillon nonetheless worried that disclosure would threaten to undermine patients' trust in the medical profession. Such commitments to absolute confidentiality were rare, however.

Risks to immediate partners were the foremost concern raised about such strong confidentiality protections. These risks were often visualized as involving "innocent" spouses who might have no idea of their partner's infidelity. As with STIs in the earlier part of the twentieth century, risk judgments about AIDS were intertwined with moral judgments about infidelity, lack of concern for the partner, and in some cases underlying ambiguities in gender identity that may have played a role in the infidelity. For example, in a case study published in 1987 in the *Hastings Center Report*, commentators debated whether a physician should tell a patient's fiancée of the risk that he was HIV positive when the patient declined to do so himself. The patient was described, somewhat pejoratively, as "bisexual," as having likely acquired the infection during one of his "homosexual encounters," and as fearful that revealing his condition would "ruin his marriage plans" (Winston and Landesman 1987, 22). Winston urged disclosure to protect the fiancée. Although Landesman judged confidentiality to be paramount, he too seriously entertained the obligation to disclose out of concern for the partner, especially when the physician did not practice in circumstances in which disclosure was likely to jeopardize his patient's trust.

Professional organizations also weighed in about partners. The American Medical Association stated that there is a duty to protect partners and recommended in an ethics opinion that if all efforts to persuade the patient to disclose failed, the physician should report the condition to authorities. If authorities failed to act, the physician should then notify and counsel the endangered partner (Lin and Liang 2005). In addition, in the Law and Medicine column of the *Journal of the American Medical Association*, Bernard Dickens (1988) outlined for physicians how law in the U.S. supported the duty to warn identified individuals at risk. Analogies were drawn to the *Tarasoff* decision (1976) in which the California Supreme Court had held that psychiatrists might have a reasonable duty to warn of their patients' immediate likelihood of violence. Raising an ethical red flag about these recommendations, one study of physicians conducted after the AMA ethics opinion indicated a greater likelihood that physicians would breach confidentiality if the patient were Black than if the patient were white (Schwartzbaum et al. 1990).

The risk of abuse was another critical concern raised about direct partner notification, particularly when it suggested that the partner might have had sex with, or shared needles with, others. Data in the U.S. suggested that HIV positive minority women in particular were at risk of domestic violence from their partners. Public health scholars argued that programs should screen for domestic violence before discussing with their female patients whether their partners should be informed about their infections (Kass and Gielen 1998). Scholars also contended that pre- and post-test counseling should include a safety plan for women and that given the low frequency of female-to-male transmission safety of the woman should override duties to the partner when the risk of violence was high. (North and Rothenberg 1993). Many states passed laws requiring partner notification, even without the consent of the patient; legal scholar Karen Rothenberg urged repeal of these statutes because of the risks of notification to women (Rothenberg and Paskey 1995).

The standard methodology of contact tracing does not involve direct notification of partners by physicians, however. Instead, physicians are required by law to report positive test results to the relevant public health authorities. To some extent, this relieves physicians of direct responsibility for any duty to reveal. Physicians can explain to their patients that they are required by law to report the information rather than seeming to make the choice themselves. Public health then makes confidential contact with the person who had the positive test result and asks him or her to name contacts who might have been exposed. This does give the person who is the identified index case some protection—he or she may be able to conceal contacts, will know that any named contacts are likely to be reached by public health, and will be assured that the identity of the possible source of the exposure will not be revealed—but concealment may be discouraged or illegal. In some states, concealment of the identity of contacts may subject the index case to criminal liability. All states receiving Ryan White funds must make good faith efforts to notify spouses of anyone reported to the public health department with a positive HIV test (HIV.gov 2018).

Even when the partner notification comes through professional public health methods, it may have serious implications for relationships. When partners are identified as positive, and their spouses or other sexual partners are notified, the information may reveal characteristics of the partner previously unknown to the other. Efforts to conceal the identity of the possible exposure source may be impractical if the partner has been in a monogamous sexual relationship. Contact tracing in such cases may be how some women become aware of their husband's bisexuality or homosexuality. It may also be how one partner learns of drug use by the other. This information may be met with grief, rejection, or anger—and, as discussed above, with psychological or physical abuse.

Challenges to the efficacy of these partner notification programs are significant. Health departments may lack resources, may be unable to find index cases or their contacts, and may not be able to obtain relevant information from index cases (Magaziner et al. 2018). Incomplete partner notification is more likely for men having sex with men, for Blacks, and for people who engage in risk behavior such as anonymous sex (Edelman et al. 2014). Given these challenges, different strategies may be preferable. One strategy is "enhanced partner notification," counseling index cases about how to communicate their status and providing them with support when they do. Another strategy is patient-delivered partner therapy, which provides index cases with the means for their partners to be treated without the need for a separate medical examination. Success of these alternatives is reportedly mixed as well and may depend on the STI in question and the context, according to a Cochrane report about the evidence from studies in many locations across the globe (Ferreira et al. 2013). Even in the U.S., domestic violence remains a significant concern for patients faced with any of these strategies and may continue to limit their use (John et al. 2018).

Other critics of these current strategies for management of the HIV epidemic urge the importance of addressing social and economic contexts for disease prevention. Efforts to identify individuals judged responsible for disease transmission and follow and control their behavior, these critics say, will fail to target environments vulnerable to HIV spread (Kazanjian 2012). They may also fail to provide the information and support needed to encourage people to seek out testing and to share information about their HIV status with their partners.

3.5.3 HIV Today

The first anti-retroviral effective for treating HIV, AZT (zinovudine) was approved in the U.S. in 1987. Today, combination therapies have turned HIV into a manageable chronic disease for most although not for all. Successful treatment also reduces viral load to levels that make transmission very unlikely. PrEP—preexposure prophylaxis—is now also available that can provide largely effective prevention for people engaging in high risk activities.

These developments have brought changes to the balance in public health recommendations about consent to testing. In 2006, the CDC issued revised recommendations for HIV testing that characterized it as routine. "Routine" means that testing is presented as part of standardly indicated medical care instead of only after an extensive structure of consent and counseling. Patients may opt out, but the expectation is that far fewer are likely to do so. The CDC recommendations were that for all patients, HIV screening should be performed at least once and that persons at high risk for infection should be screened annually. For pregnant women, the recommendation was also that screening should be included in prenatal blood tests and should be repeated in the third trimester in areas of high rates of HIV infection among pregnant women (Branson et al. 2006). This transition was defended ethically because of the availability of treatment and the role of treatment in prevention but generated ethical controversy as well. Although today HIV is more of a chronic disease than an inevitably fatal one, knowledge of their HIV status continues to subject people to risks. As just described, the majority of U.S. jurisdictions still criminalize knowing transmission of HIV. People with HIV may have difficulty with employment or insurance, may face rejection from family and friends, and may be subject to violence from their intimate partners. These risks were higher when compounded by other factors of disadvantage such as race or poverty. These risks also were more severe at the outset of the HIV epidemic, so it should come as no surprise that AIDS advocacy emphasized the importance of informed consent to testing despite the possibility that it would discourage individuals from knowledge of their HIV status and reduce the ability of public health to deal successfully with the pandemic.

Still, these therapies are expensive and not always effective. Pre-exposure prophylaxis (PrEP) has also become available for those engaging in risky behavior and seeking to reduce their risks of infection. However, a continuing course of PrEP costs about $2000 per month in the U.S. and is not always covered by health insurance plans; moreover, people taking PrEP may be denied disability or life insurance (McNeil 2018). The evidence is that PrEP use is far less frequent among African American and Latino patients—groups in which HIV incidence remains high (CDC 2018). Data from trials in France and in Canada suggest that on demand PrEP may be far less expensive, receive better patient adherence, and be highly successful in preventing HIV transmission (Molina et al. 2017).

According to UNAIDS (2018), the agency of the United Nations founded in 1996 to fight spread of the disease, 78 million people worldwide had become infected by AIDS by 2018 and over 35 million had died. UNAIDS began pursuing a fast track strategy in 2014 aimed to bring the AIDS pandemic under control by 2030, with control defined as 90-90-90—a 90% rate of diagnosis of those infected, a 90% rate of treatment of those diagnosed, and a 90% rate of viral suppression in those undergoing treatment (UNAIDS 2018a). As of the halfway point to the 2020 benchmarks, the program had fallen behind in some areas of major concern. Progress has been particularly good in areas of eastern and central Africa hard hit by the epidemic, with significant increases in treatment rates and concomitant declines in mortality. Although rates of new infection fell significantly in areas with effective access to treatment and PrEP, rates remained high in other areas. Rates of untreated children remained particularly high. Declines in new infections were lowest in the Americas and new infection rates rose significantly in the Middle East and Eastern Europe. In addition, there are fears that resistant forms of HIV are appearing that may prove difficult to treat (Garrett 2018). Surveillance remains critically important to monitor these developments.

3.6 Ethical Tensions

Case finding and contact tracing bring potential tensions between the individual and the overall social good into sharp focus. On the one hand, respect for individual rights may seem to support the importance of informed consent, confidentiality, and freedom to choose whether to be tested for a disease or to have test results used to protect others. It may also seem to favor rights to gain swift access to potentially effective medications even before testing for safety and efficacy are complete. Respect for individual rights also supports protection of people in research, even when that might slow the development of useful knowledge for the communities in which they live. On the other hand, choices made by individuals may put others at risk, such as their partners, the children they bear, or the patients they treat. Case identification and contact tracing may be necessary and effective ways to reduce disease spread. Yet if use of these methods discourages people from any encounters with health care at all, the result may backfire not only for the disease in question but for other aspects of health as well. When prevention or treatment are readily available, direct efforts to educate people about a disease and what might be available to them could prove more beneficial for individuals and more effective in reducing disease altogether than any efforts to trace individual chains of disease from an index case. These are the apparent tensions between respect for the individual and protection of others who may be harmed. We say "apparent," however, because these tensions emerge most forcefully when autonomy is understood in a particular, individualistic way.

3.6.1   Individualistic Autonomy and Informed Consent

The field of bioethics, maturing just as the AIDS crisis flowered, brought individual autonomy to the fore of medical decision making. The initial edition of the *Principles of Biomedical Ethics*, by Beauchamp and Childress, was published in 1979, and has been a core reference text in the field through multiple editions ever since. The volume argued for what became the classic litany of bioethics principles: autonomy, beneficence, non-maleficence, and justice. Autonomy reflected the ideas of dignity, integrity, and self-determination that had been critical in the Declaration of Helsinki adopted in 1964 by the World Medical Association (WMA) to protect human subjects in research (2019). The WMA declaration, spurred by the recognition of research atrocities not only in Nazi Germany but also in the U.S. and elsewhere, insisted on free and voluntary informed consent as an ethical condition of participation in research studies.

Use of the term "autonomy" as the ethical underpinning for informed consent had overtones beyond voluntariness and full information. The idea of autonomy was drawn from the Kantian vision of people as individual rational agents. As such, it was rooted both in rationality and in individualism. Freed from the constraints of their desires and interests, Kantian individuals laid down the moral law for themselves, subject only to the constraints of universal rationality. As such rational beings, willing universally, they would will that they treat others as they would want to be treated themselves, free of the bonds of emotions or desires or presupposed connections to others.

Such Kantian autonomy has been vigorously criticized by feminists and communitarians. Feminists developed views of the importance of relationships to identity, arguing that individuals are not metaphysically separate selves but selves-in-relationships to others. People are mothers or fathers, friends or enemies, Christians or Jews, Americans or French or Chinese. Oppressive relationships are problematic, but relationships themselves—to friends, families, communities, countries, or even the natural world—are not. Communitarians argued further, and controversially, that any individual choices take place in the context of community ties and values. Although communities may be oppressive and people should be free to leave—as Amy Gutmann (1985, 319) famously pointed out, in Salem they felt compelled to burn witches—the shared values of community are part of the moral landscape and perhaps even take precedence over individual values and interests. These disputes are of course far more complex than a brief, caricature-like sketch, can give, and we explore them further in Chapter 7. But it is important to recognize that much of the early bioethics support for informed consent was rooted in individualism.

The models of consent thus employed in bioethics started with individuals' values and preferences, not with relationships or communities. Individuals should understand their conditions, the treatment options available to them (including no treatment), and the risks, benefits, and likely outcomes of each of these treatment options with respect to their values. Consenting individuals were supposed to be able to deliberate in a way that would enable them to apply their knowledge to choose the alternative that best reflected their values. People with cognitive impairments or mental illness might be capable of giving informed consent in this way, although they might need extra support to do so. So-called "vulnerable" populations—children, pregnant women, prisoners—or people who have suffered disadvantage or discrimination might also require special attention to ensure their informed consent.

Discussions in bioethics emphasized this model of informed consent in both clinical care and research regarding HIV/AIDS. In clinical care, advocates argued for a full consent process prior to HIV testing and for pre- and post-test counseling. Recommendations for the consent process included explanations of the potential personal and social risks of testing. Counseling was to help people deal with the consequences of test results and might include both psychological support and referrals for needed social services. For example, Carol Levine and Ron Bayer argued in the *American Journal of Public Health* that despite the possibility that early intervention might delay the onset of disease, screening asymptomatic individuals in high risk populations must be fully voluntary and take place only after careful informed consent. They wrote: "It is precisely when medicine's capacity to enhance patient welfare appears to be increasing that there is a danger that important ethical concerns can be overridden or disregarded. This is especially so in the case of AIDS ..." (1989, 1661).

This consent model for HIV testing gained support in public policy as well. Laws were enacted requiring informed consent prior to HIV testing and protecting the confidentiality of test results to encourage people to learn their HIV status. The American Hospital Association recommended to its members that general consents to blood testing obtained on admission to the hospital did not include HIV testing and that such testing solely for the protection of health care workers should be discouraged, as these workers could instead be expected to avoid exposure by the use of CDC-recommended universal precautions such as gloves and masks (Swartz 1987).

Informed consent mandates for HIV testing were not without critics, however. In 1987 members of the British Medical Association voted that HIV testing should be at the physician's discretion acting in the patient's best interest and should not necessarily require consent. The BMA vote echoed the sentiments of a dermatologist who wrote to the *British Medical Journal* that he should be able to test to rule out HIV as a cause of eczema without alarming a patient who was about to be married. This physician's letter protested the "almost hysterical approach" of the British National Health Service in requiring consent for HIV testing (Shrank 1987). The British Medical Association's lawyers soon spoke out, however, publishing guidance that such testing might subject physicians to criminal liability for battery or to civil liability for negligent failure to obtain informed consent (BMA 1987). Other commentators questioned "AIDS exceptionalism" and argued that there was no apparent difference between HIV testing and other blood tests with respect to consent (e.g. Cohen 1988).

3.6.2 Access to Experimental Drugs for HIV

Autonomy was also used to support AIDS patients' efforts to access drugs that might be useful in treating their disease. At the outset, AIDS was a fatal disease and patients argued that they should be at liberty to try therapies with a chance of working even though the safety and efficacy of these therapies had not been fully investigated. Access to treatment could also encourage testing and identification of persons with HIV infection. In response to what were seen as sympathetic appeals on behalf of desperately ill patients and FDA bureaucracy and delays, a parallel fast track was developed for approval of drugs to treat HIV. The fast track was controversial, especially after patients receiving early access to the investigational drug dideoxyinosine died at rates higher than expected. In response, the Institute of Medicine (IOM) convened a conference to discuss the controversy and pointed out the importance of continued ethical monitoring of the early access program, its impact on patients, and its consequences for the development of effective treatments (IOM 1990). The IOM criticized what it regarded as extreme arguments that because patients have little to lose their access to drugs should not be restricted for their failure "to recognize the association between desperation and vulnerability … People with HIV infection do have something to lose: they can waste time, energy, and hope—or even become sicker—on substances that would never reach the marketplace through normal channels" (IOM 1990 Ch. 2) The report also pointed out that without funding for health care, patients with HIV might be dependent on participation in drug trials in order to receive any care at all.

Another set of criticisms of the fast track relied on the possibility of deleterious consequences for others. One problem was that the rush to access might undermine participation in the clinical trials needed to gain adequate information about the safety and efficacy of novel therapies. Relatedly, uncontrolled use of experimental therapies might result in unrepresentative adverse events that could lead to premature rejection of a promising opportunity. Or, it might generate unrepresentative success and resulting premature optimism leading to widespread but fruitless use. Widespread use of novel therapies might have more general population effects as well, if people believe that their disease has been sufficiently managed and engage in risky behaviors, or if people use the therapy unevenly and resistant strains of infection develop. Advocacy on behalf of HIV-positive patients for access to experimental therapies has developed into a more general movement for "right to try" laws, now adopted in the vast majority of U.S. states and by the U.S. federal government (Joffe and Lynch 2018).

For surveillance, one potentially problematic result of "right to try" is that use of new therapies may become dispersed without systematic collection of information about the consequences. If novel therapies move into clinical practice without monitoring, information may never be developed about effects on subpopulations, either beneficial or deleterious, or rarer side effects and how to recognize or manage them. Needs for such knowledge about patients' experiences with untested therapies may provide arguments for patients to share information and for increased surveillance to collect it. On the other hand, if "right to try" is coupled with ongoing collection of information about patient outcomes and if the possibility of gaining early access to care encourages participation, the consequences for surveillance may improve.

3.6.3 Research Ethics and HIV

Concerns about informed consent were especially intense when clinical trials about HIV were conducted in developing countries that could not (or did not) afford the costs of HIV care. The effects of these trials continue to resonate for surveillance, not always well. Large geographical areas of central and southern Africa were hit particularly hard by the HIV pandemic. Clinical trials were designed to test whether in these circumstances short-course anti-retroviral therapy in the perinatal period could reduce the risks of maternal-fetal transmission of disease. Critics of the trials were outraged because the research compared short-course anti-retroviral therapy just before birth to placebo—no treatment—rather than to the better treatment that had been offered when similar trials were conducted in the U.S. (Lurie and Wolfe 1997).

The justification for the trials was that no treatment was all that was available for the participants without the trial. Moreover, the only likely treatment that might become available in their circumstances was the short-term therapy, and it was important to see whether even that could make a difference in the conditions in which it might eventually be implemented. Defenders contended that the ethical permissibility of the trials should be assessed against the background of the locations in which they were conducted, and that it was thus unfair to compare the placebo-controlled AIDS trials to the Tuskegee syphilis study in the U.S. where treatment was available and had been deliberately withheld from participants for many years (Varmus and Satcher 1997).

Nonetheless, the critics contended that researchers should be held to the same standards whether they were conducting a study in the U.S. or elsewhere. Research is not clinical care, so, it is inappropriate to rely on the availability of clinical care to assess the ethics of the research. Critics also argued that informed consent to participation in the trials was deeply problematic, as participants lacked the education and experience to understand what might be happening or the resources to be able to consider alternatives freely. Moreover, critics argued, the idea of individualized informed consent was alien to the communities in question, where models of decision-making were communal rather than individual. To meet this concern, some researchers explored communal models of consent such as the traditional community forums (*mabaraza*) used in East Africa (Vreeman et al. 2012), models that we explore further in Chapter 7. Others argued that researchers had obligations to provide participants or others in their community with continuing effective treatment or other benefits such as improved access to primary care (Shaffer et al. 2006). Marcia Angell (2000), then editor of the *New England Journal of Medicine*, explained a decision to publish one of these controversial studies by saying that although she thought the researchers' obligations should not vary with the context, others judged the research to be acceptable because of its potential benefit.

Another example of controversial research in Africa also has resulted in continuing problems for public health. During a meningitis epidemic in Nigeria in 1996, Pfizer conducted a clinical trial of its experimental antibiotic Trovan. Charges about the trial included that it had not been property approved by the Nigerian authorities, that participants had not given informed consent, that the drug was inappropriate for use in sick children, and that public health personnel were diverted from public health to the Pfizer research (Ezeome and Simon 2010). The trial became the subject of *The Constant Gardener*, a movie dramatizing exploitation by pharmaceutical companies. Pfizer eventually paid compensation to the families of four children (Smith 2011). In the aftermath of the trial, critics urge higher standards of informed consent for global medical research (e.g. Annas 2009). Memories of the trial may be part of the explanation for suspicions about public health efforts to encourage polio vaccination in the affected region of Nigeria (Closser et al. 2015).

Similar issues about informed consent and the ethics of research may also come into play where information is gathered for public health purposes. The line between "research" and "public health" may itself be obscure, if both gather data to produce generalized knowledge. This blurring suggests the importance of ensuring that information gained by public health is used for health.

3.7   COVID-19 and Enhanced Contact Tracing

The COVID-19 pandemic has features that are especially challenging for contact tracing (Moore et al. 2020). Because people may be highly contagious before any symptoms appear—or even without ever having symptoms—index cases may come to light too late to prevent wide community spread. Possibilities of spread through aerosolized droplets and fomites mean that people may become infected without ever knowing about their contacts. In some communities, mistrust of public health and concerns about racism have dampened willingness to share information. Nonetheless, information about the whereabouts and possible contacts of infected individuals, especially when they were asymptomatic or pre-symptomatic, became critical to the effort to stop the spread of this highly contagious disease.

Different countries took different routes towards enhancing contact tracing to address these COVID-19 challenges. Some in addition used smart technology to monitor whether people were conforming with orders to remain sequestered to avoid risks of their spreading disease. These enhancements raised significant concerns about privacy and coercion and met with mixed success at least during the initial six months of the pandemic. It is fair to say that the more intrusive the methods the more they appear to have contributed to pandemic control, although other factors may have been at work as well. Here, we discuss ethical questions raised by these methods as they contributed to contact tracing. In Chapter 5, we consider further issues raised by novel data sources and uses, some of which have been used in combination with enhanced contact tracing methods.

China used a mandatory approach to collecting information and monitoring behavior, according to descriptions available from news reports. Everyone is required to use software on their smart phones to indicate their contagion risk. The system generates a color code—red, yellow, or green—for each user. The color code is based on personal integrity reporting and comparisons with information held by the government (Xinhuanet 2020). The code is dynamic, updated with information about the user's own status and status of the area in which the user has been. Code status is checked upon entry to places such as shopping malls or subway stations; people without their codes or who are not "green today" are turned away. Yellow status means continued self-quarantine for at least seven days and red status for up to fourteen. According to reports, the system was developed by Chinese internet companies working in tandem with the government and police may be notified of individuals' locations (Mozur et al. 2020).

Reports have surfaced that use of the method is not limited to COVID-19 surveillance. One Communist Party official reportedly stated that the tracking app should be an "intimate health guardian" that is "loved so much that you cannot bear to part with it" (Zhong 2020). Reports also suggest that data collected by the apps may be used for purposes other than health, including policing. While China has reportedly been very effective in stopping the outbreak, and in responding to new bursts of infection, its method allows for other forms of surveillance that may significantly limit individual liberty and result in other harms to individuals.

In its efforts to trace cases of COVID-19, South Korea also used a non-voluntary method that centralized data compilation. In contrast to China, South Korea limited any use of the data it collected to COVID-19 surveillance. When patients were confirmed with COVID-19, South Korea compiled a dossier that was designed to provide as complete as possible a picture of where they might have been. Dossiers included smartphone data, credit card transactions, immigration information, and footage from security cameras (Jo 2020). These dossiers were then used to identify and notify possible contacts as well as to identify hotspots such as bars where a number of transmissions might have taken place (Lin and Martin 2020). Information about virus hotspots was publicized so that people who might have been at the scene could learn about their possible risks. Although individual information was only released by number rather than by personal identifiers, some people were reidentified by their movement histories and "doxed" or otherwise stigmatized (Jo 2020; Kim 2020). After the South Korean Human Rights Commission urged further privacy protections, the guidelines were amended to exclude disclosure of any information that could be personally identifiable (Jo 2020). However, routes that a confirmed-positive patient traveled could still be disclosed, leading to ongoing concerns about individual privacy and the consequences for businesses named as sites visited by infected persons (Kim 2020). In addition, people who are isolated due to exposures have apps that track their whereabouts to make sure they are adhering to the restrictions. Importantly, none of the information gathered in Korea for COVID-19 surveillance could be used for any purpose other than disease outbreaks, including public safety or national security. One commentator suggests this might be called "virtuous surveillance—a radically transparent version of people-tracking that is subject to public scrutiny and paired with stringent legal safeguards against abuse" (Kim 2020).

In cooperation with technology companies, a variety of groups have been developing more privacy-protective ways to enhance contact tracing by using smart devices. These methods differ by whether they rely on centrally stored data or data stored only on individual phones and by the extent of user control. Data storage on individual phones is the most privacy protective. However, evidence as of this writing suggests that it is the least useful—although this may change if people perceive this method as trustworthy and effective and thus are encouraged to sign up for it.

One technology uses mobile location data to trace where an individual has been. Individuals carrying their cellphones may be traced by triangulating cellphone towers with which a user's phone has been in contact to attempt to determine their precise locations. The information is not stored on the individual's phone but is stored by the telecommunications companies. Another location tracing technology uses GPS (global positioning system) information collected by apps running on cellphones. Mapping apps, fitness apps, or apps recommending nearby restaurants collect this information. It is often harvested for commercial purposes such as advertising or risk assessment based on knowledge of the app user's habits. Such locational information may be highly sensitive, revealing that an individual has entered a cancer treatment center, an abortion clinic, a cannabis dispensary, or a shop selling sex toys, among many other possibilities. Governments may also seek access to the information to determine the individual's whereabouts (as indicated by the cellphone location or from the apps); in 2018, the US Supreme Court held that month-long collection of locational data from cellphone records without a search warrant violated the user's right to privacy (*Carpenter* 2018).

Locational data from cellphones can be used to trace the movements of individuals who have been diagnosed with COVID-19. Because it is time-stamped, it can also identify individuals who may have been in the same location for times when

exposure might have been possible. Geolocational tracking may help to jog individuals' memories about places they visited while they may have been contagious
and who they may have encountered during these visits. Privacy advocates and
advocates of human rights have raised concerns about these methods, however.
Unless individuals have the ability to abandon their phones when they go to sensitive locations, turn off any geolocation functionality, or consent to use of this data
for contact tracing, use of the information will not be with their consent or under
their control. Critics are also very concerned that if individual geolocation data
becomes available to public health it may be used by police, security, or immigration authorities to track movements. Human Rights Watch (2020) has identified
some emergency decrees for use of geolocation data in COVID-19 tracking that
may lead to human rights abuses. Privacy concerns led the state of Utah to turn off
the locational tracking function in its COVID-19 app, Healthy Together, leaving the
app without one of its initial selling points (Rodgers 2020).

   Bluetooth technology is more protective of individual privacy. This technology—
used for example to enable wireless linkage between cellphones and earphones—
picks up signals indicating proximity with other similarly equipped devices. Use of
Bluetooth does not require locational information although the two can be combined. Data can be stored on individual devices rather than centrally collected. A
partnership between Apple and Google has built an infrastructure that allows individuals to download an app that creates an encrypted identifier for their device
(Nield 2020). That app will record the identifiers for any devices also using the app
that are in sufficiently close proximity to be picked up by Bluetooth—a distance up
to about 30 feet (Greenberg 2020). An individual who tests positive for COVID-19
and who also has the app can then send a signal to other app users that lets them
know that a device with the individual's encrypted identifier has been close to them,
and they should take steps to be tested and to isolate. As originally described, the app
did not share either locational information or the information about the potential
exposure source—just the fact that there was device proximity sufficient to suggest
the possibility of exposure. The approach seemed more likely to be attractive to
people with strong privacy protective concerns. However, reports surfaced in July
2020 that when used on Android systems Google might pick up locational information from some users and countries using the app sought clarification (Singer 2020).
The disadvantage of this technology is its limited efficacy and the possibility that it
may create a false sense of security if people believe it will reveal any exposures
they might have experienced (Cohen et al. 2020). Without locational information, it
will not identify devices that passed through the area at an earlier or later time but
that could be an exposure source. Moreover, it only works if a sufficient number of
people voluntarily download the app and enter any positive test results into it.

   Indeed, the COVIDSafe app used by Australia is an apt illustration of the difficulties in rolling out this Bluetooth approach. Individuals re invited to download the
app and provide their name, mobile number, postcode, and age range (Australia
Health 2020). They then receive a confirming text message to complete the installa-
tion. An encrypted reference code is created as a unique identifier. Then, as app

users go about in the world, they must keep their phone running with the app turned on and Bluetooth enabled. Whenever a phone running the app gets sufficiently close to another user's phone, it notes the date, time, distance and duration of the contact, along with the encrypted reference code of that user, but not the location of the contact. The information, stored only in the app on the user's phone, is automatically fully deleted after 21 days, the outside window of possible infection. If a user tests positive, a public health contact tracer contacts the user in the usual way. The interviewer asks whether the user has the app and is willing to allow data from the app to be used in contact tracing—another point at which the user has the option whether to participate. If the user agrees, encrypted information from the app is uploaded to a secure information storage system. Health officials then use the information to identify contacts and proceed in the ordinary way for contact tracing. The only difference between this method and contact tracing that is not digitally enhanced is the information provided by the app about proximity with other users.

Australia was initially very optimistic about the potential for the app. However, several months into the pandemic, it appeared that the app had not added very much to Australia's success in pandemic control, possibly identifying at most one case of infection that would not have been caught through ordinary contact tracing methods (Taylor 2020). One problem was user uptake. Australia set a target of about 40% of the population using the app but appears to have fallen about 1.5 million users short of that target. Some Australians do not have smartphones, others use them inconsistently, and others were unable or uninterested in downloading the app. Poll data indicated that Australians did not think that the app would be particularly effective or that the threat of the virus was sufficiently great to warrant use of the app.

Another more significant problem with COVIDSafe was an initial difficulty for use of the app on Apple iPhones. Apple privacy protections required that the app run on the screen of the phone, an impractical requirement for most users because of battery use. The Apple-Google interface moves the Bluetooth connection process to the operating system but will not give identifying information about the phones with which the user has been in contact. Instead, it will send alerts to other users that their phones have had a potential exposure. It is then up to these other users to follow up and get tested. The COVIDSafe app would need to be retrofitted to employ this functionality.

Other countries rolled out apps that incorporated the Apple-Google interface. Latvia launched its app, Apturi (Stop) Covid, at the end of May 2020. Latvia's app was designed to work within the country but with the possibility of interfacing with other European countries such as Germany (Reuters 2020).

Digitally enhanced contact tracing thus presents an informative example of the balance between surveillance efficacy and protection of what many see as individual rights. Methods that are most effective are not very protective of individuals, especially if they are not voluntary, store information centrally, and allow information to be used for purposes other than pandemic tracing. China illustrates perhaps the most problematic version of these methods for individuals as victims; South Korea's method is more protective but still includes some secondary risks. Methods that are very privacy protective are unlikely to add much to traditional contact tracing, as the Australian example indicates. Making the methods non-voluntary and adding geolocation data to them greatly increases efficacy. But it also emphasizes the importance of transparency and controls on data use other than for pandemic surveillance.

3.8   Informing the Subjects of Reports

Laws require many contagious diseases to be reported to public health. Reporting requirements may extend beyond efforts to control contagion through case finding and contact tracing. In Chapter 5, we consider some of these other data collections by public health, including disease registries and tracking the incidence of noncontagious conditions such as cancer or birth defects. Here, we focus primarily on reports of contagious diseases that track disease spread and are used for contact tracing. Unless their providers tell them, patients may be unaware that reporting is taking place. This question of notice presents conflicting risks for surveillance.

On one side, some fear that informing patients about reports may increase reluctance to seek care and reduce patients' trust in how their physicians manage their medical records. People who know their contagion has been reported may also seek to evade public health officials. Public suspicion may be aroused by the knowledge that governmental entities possess information, even when that information has been deidentified and is not shared for the purpose of tracking contagion. Moreover, informing patients requires an additional step on the part of the physician that may result in patient requests for further explanations and time-consuming reassurance, although the burdens of notification can be reduced with electronic communication. Protection of those who make reports, who may not always be physicians, is also a concern. Finally, as we discuss in detail in Chapter 5, state law protections for information possessed by public health are uneven.

On the other side, there are significant advantages to informing people that they have been the subject of reports to public health. Mistrust may result if people learn that data have been shared without their knowledge. Transparency is often a starting point for trust, as the lack of transparency is for mistrust. Knowledge that the information has been revealed can enable people to take steps to protect themselves if further disclosures occur. The case for informing people that information has been shared is especially strong when public health data safeguards are relatively weak. This allows people to take steps to protect themselves such as informing others of their conditions in a way that is designed to allay concerns and mitigate unfavorable reactions to the extent possible. When information about data sharing and use is public, moreover, watchdog organizations may also act to protect the public or groups within the public from data misuse.

Informing people is a way of letting them know about the possible benefits of public health, too. People can be told about the knowledge gained from shared data for themselves and for their community. Information about public health data use may include information for patients about treatment that may be available for their conditions. Regular updates can tell communities what is being accomplished with public health data and how it contributes to improving health. These factors, we think, will in most contemporary circumstances weigh in favor of routinely informing individuals and the public when data have been shared with public health. The more that people recognize the benefits of public health data use for themselves and their community, and the more they receive information and support for their own health, the more sustainable surveillance will be.

3.9 Summary

Identification of index cases, in combination with tracing their contacts, remains a commonly recommended form of surveillance to increase treatment frequency and reduce disease spread. When chains of transmission can be followed easily, its efficacy is greatest. It may be especially helpful in situations in which a new disease is introduced into a community in which it was not already widespread. Its benefits are clearest, and risks reduced, when effective treatment is readily available.

Case identification and contact tracing may cause significant harm to those it implicates, however. Historically these methods have been associated with judgments of moral condemnation, particularly of individuals judged to be the sources of sexually transmitted diseases. These risks were especially clear at the outset of the HIV/AIDS epidemic, as the disease was deadly, untreatable, and its modes of transmission were poorly understood. Because of these risks to individuals, testing for HIV/AIDS initially required elaborate informed consent. Controversies arose about whether patient confidentiality ought to be protected. Arguments for breaching confidentiality were judged to be strongest when unknowing partners were at immediate risk, but with the recognition that relationships could be forever altered and individuals' personal and economic security severely threatened. Panic about the possibility that health care providers could transmit disease—or that disease might otherwise be transmitted in workplaces—gradually subsided in the face of evidence that the initial panic was unwarranted.

Case identification and contact tracing therefore raise a balance of ethical considerations which may weigh differently depending on the context. They are most morally compelling when their benefits can be maximized, their harms can be minimized, and better alternatives for disease prevention are lacking. In the best of circumstances, when persons who have become infected can be identified without stigma, and treated without shame, they are an ethically sustainable form of public health surveillance.

References

Adler, Michael W. 1980. The Terrible Peril: A Historical Perspective on the Venereal Diseases. *British Medical Journal* 281: 206–211.

Alexander, Thomas S. 2016. Human Immunodeficiency Virus Diagnostic Testing: 30 Years of Evolution. *Clinical and Vaccine Immunology* 23 (4): 249–253.

Anderson, Tim. 2009. HIV/AIDS in Cuba: A Rights-Based Analysis. *Health and Human Rights* 11 (1): 93–104.

Angell, Marcia. 2000. Investigators' Responsibilities for Human Subjects in Developing Countries. *New England Journal of Medicine* 342: 967–969.

Annas, George. 2009. Globalized Clinical Trials and Informed Consent. *New England Journal of Medicine* 360: 2050 2053.

Armbruster, Benjamin, and Margaret Brandeau. 2007. Contact Tracing to Control Infectious Disease: When Enough is Enough. *Health Care Management Science* 10 (4): 341–355.

Australian Government Department of Health (Australia Health). 2020. COVIDSafe app. https://www.health.gov.au/resources/apps-and-tools/covidsafe-app#:~:text=The%20COVIDSafe%20app%20helps%20find,with%20someone%20with%20COVID%2D19. Accessed 15 July 2020.

Bayer, Ronald, and Cheryl Healton. 1989. Controlling AIDS in Cuba: The Logic of Quarantine. *New England Journal of Medicine* 320: 1022–1024.

Beauchamp, Tom L., and James F. Childress. 1979. *The Principles of Biomedical Ethics*. New York: Oxford University Press.

Boone, Stephanie A., and Charles P. Gerba. 2007. Significance of Fomites in the Spread of Respiratory and Enteric Viral Disease. *Applied and Environmental Microbiology* 73 (6): 1687–1696.

Brandt, Allan M. 1985. *No Magic Bullet: A Social History of Venereal Disease in the United States Since 1880*. New York: Oxford University Press.

Branson, Bernard M., H. Hunter Handsfield, Margaret A. Lampe, Robert S. Janssen, Allan W. Taylor, Sheryl B. Lyss, and Jill E. Clark. 2006. Revised Recommendations for HIV Testing of Adults, Adolescents, and Pregnant Women in Health-Care Settings. *Mortality and Morbidity Weekly Report* (Sept. 22)/55(RR14): 1–17.

British Medical Association (BMA). 1987. HIV Antibody Testing: Summary of BMA Guidance. *British Medical Journal* 295 (10 Oct): 940.

Carlisle, Nate. 2019. Utah's Controlled Substance Database Susceptible to Hacking, Auditors Warn. Salt Lake Tribune [online] (Feb. 13). https://www.sltrib.com/news/politics/2019/02/13/utahs-controlled/. Accessed 15 July 2020.

Castro, Arachu, and Paul Farmer. 2005. Understanding and addressing AIDS-related stigma: from anthropological theory to clinical practice in Haiti. *American Journal of Public Health* 95 (1): 53–59.

Carpenter v. United States, 585 U.S. ___, 138 S.Ct. 2206 (2018).

Centers for Disease Control & Prevention (CDC). 2019. National Notifiable Diseases Surveillance System (NNDSS). https://wwwn.cdc.gov/nndss/. Accessed 15 July 2020.

———. 2018. HIV Prevention Pill Not Reaching Most Americans Who Could Benefit—Especially People of Color (Mar. 6). https://www.cdc.gov/nchhstp/newsroom/2018/croi-2018-PrEP-press-release.html. Accessed 15 July 2020.

Closser, Svea, Anat Rosenthal, Kenneth Maes, Judith Justice, Kelly Cox, Patricia A. Omidian, Ismaila Zango Mohammed, Aminu Mohammed Dukku, Adam D. Koon, and Laetitia Nyirazinoyoye. 2015. The Global Context of Vaccine Refusal: Insights from a Systematic Comparative Ethnography of the Global Polio Eradication Initiative. *Medical Anthropology Quarterly* 30 (3): 321–341.

Cohen, Carl. 1988. Commentary on Case Studies: What is the Difference Between and HIV and a CBA? *The Hastings Center Report* 18 (4): 19–20.

Cohen, I. Glenn, Lawrence O. Gostin, and Daniel J. Weitzner. 2020. Digital Smartphone Tracking for COVID-19-Public Health and Civil Liberties in Tension. *Journal of the American Medical Association* 323 (23): 2371–2372.

Committee for the Study of the Future of Public Health (Committee). 1988. *The Future of Public Health*. Washington, DC: National Academy Press.

Cooper, Elizabeth B. 2001. Social Risk and the Transformation of Public Health Law: Lessons from the Plague Years. *Iowa Law Review* 86: 869–947.

Department of Health and Human Services (DHHS). 1979. *Healthy People: The Surgeon General's Report on Health Promotion And Disease Prevention*. https://profiles.nlm.nih.gov/spotlight/nn/catalog/nlm:nlmuid-101584932X94-doc. Accessed 15 July 2020.

Deporte, J.V. 1941. Premarital and Prenatal Tests for Syphilis. *The Lancet* 238 (6150): 59.

Dickens, Bernard. 1988. Legal Limits of AIDS Confidentiality. *Journal of the American Medical Association* 259 (23): 3449–3451.

Dillard v. O'Kelley, 961 F.3d 1048 (8th Cir. 2020).

Doe v. Marselle, 675 A.2d 835 (Conn. 1996).

Doe v. University of Maryland Medical System Corporation, 50 F.3d 1261 (4th Cir. 1995).

Edelman, E.J., K.S. Gordon, M. Hogben, S. Crystal, K. Bryant, A.C. Justice, and D.A. Fiellin. 2014. Sexual Partner Notification of HIV Infection Among a National United States-Based Sample of HIV-Infected Men. *AIDS & Behavior* 18 (1): 1898–1903.

Evans, Erica. 2017. How a 2015 Law Change Affected Law Enforcement's Fight Against the Opioid Crisis. *Deseret News* [online] (Dec. 6). https://www.deseretnews.com/article/900005271/how-this-2015-law-change-affected-the-fight-against-the-opioid-crisis.html. Accessed 15 July 2020.

Ezeome, Emmanuel R., and Christian Simon. 2010. Ethical Problems in Conducting Research in Acute Epidemics: The Pfizer Meningitis Study in Nigeria as an Illustration. *Developing World Bioethics* 10 (1): 1–10.

Fallone, Edward A. 1988. Preserving the Public Health: A Proposal to Quarantine Recalcitrant AIDS Carriers. *Boston University Law Review* 68 (2): 441–506.

Fairchild, Amy L., David Rosner, James Colgrove, Ronald Bayer, and Linda P. Fried. 2010. The EXODUS of Public Health: What History Can Tell Us About the Future. *American Journal of Public Health* 100 (1): 54–63.

Fairchild, Amy L., Daniel Wolfe, James Keith Colgrove, and Ronald Bayer. 2007. *Searching Eyes: Privacy, the State, and Disease Surveillance in America*. Berkeley: University of California Press.

Fee, Elizabeth, and Manon Parry. 2008. Jonathan Mann, HIV/AIDS, and Human Rights. *Journal of Public Health Policy* 29 (1): 54–71.

Ferreira, A., T. Young, C. Mathews, M. Zunza, and N. Low. 2013. Strategies for Partner Notification for Sexually Transmitted Infections, Including HIV. Cochrane Database of Systematic Reviews 3(10): CD002843. https://www.cochranelibrary.com/cdsr/doi/10.1002/14651858.CD002843.pub2/abstract. Accessed 15 July 2020.

Frith, John. 2012. Syphilis—Its Early History and Treatment until Penicillin and the Debate on its Origins. *Journal of Military and Veterans' Health* 20 (4): 49–58.

Gabriel, Trip. 2020. Ohio Lawmaker Asks Racist Question About Black People and Hand-Washing. *The New York Times* [online] (June 11). https://www.nytimes.com/2020/06/11/us/politics/steve-huffman-african-americans-coronavirus.html. Accessed 25 July 2020.

Garrett, Laurie. 2018. Welcome to the Next Deadly AIDS Pandemic. *Foreign Policy* (July 25) [online]. https://foreignpolicy.com/2018/07/25/welcome-to-the-next-deadly-aids-pandemic/?utm_source=PostUp&utm_medium=email&utm_campaign=Editors%20Picks%20%207/25/2018%20-%20BSA&utm_keyword=Editor#39;s%20Picks%20OC. Accessed 15 July 2020.

Gillon, Raanan. 1987. AIDS and Medical Confidentiality. *British Medical Journal* 294 (27 June): 1675.

Gostin, Larry. 1986. The Nucleus of a Public Health Strategy to Combat AIDS. *Law, Medicine and Health Care* 14 (5–6): 226–230.

Gostin, Lawrence O., and James G. Hodge Jr. 1998. Piercing the Veil of Secrecy in HIV/AIDS and other Sexually Transmitted Diseases: Theories of Privacy and Disclosure in Partner Notification. *Duke Journal of Gender Law & Policy* 5: 9–88.

Greenberg, Andy. 2020. How Apple and Google Are Enabling Covid-19 Contact-Tracing. *WIRED* [online] (April 10). https://www.wired.com/story/apple-google-bluetooth-contact-tracing-covid-19/. Accessed 15 July 2020.

Gruskin, Sofia. 2002. The UN General Assembly Special Session on HIV/AIDS: Were Some Lessons of the Last 20 Years Ignored? *American Journal of Public Health* 92 (3): 337–338.

———. 2002a. Ethics, Human Rights, and Public Health. *American Journal of Public Health* 92 (5): 698.

Gutmann, Amy. 1985. Review: Communitarian Critics of Liberalism. *Philosophy and Public Affairs* 14 (3): 304–322.

Health Insurance Portability and Accountability Act Rule (HIPAA), 45 C.F.R. § 164.512(b) (2019).

HIV.gov. 2018. Limits on Confidentiality: HIV Disclosure Policies and Procedures. https://www.hiv.gov/hiv-basics/living-well-with-hiv/your-legal-rights/limits-on-confidentiality. Accessed 15 July 2020.

Human Rights Watch. 2020. Mobile Location Data and Covid-19: Q&A (May 13). https://www.hrw.org/news/2020/05/13/mobile-location-data-and-covid-19-qa. Accessed 15 July 2020.

Institute of Medicine (IOM). 1990. *Expanding Access to Investigational Therapies for HIV Infection and AIDS: March 12–13, 1990 Conference Summary*. Washington, DC: National Academies Press.

John, Steven A., Jennifer L. Walsh, Young If Cho, and Lance S. Weinhardt. 2018. Perceived Risk of Intimate Partner Violence Among STI Clinic Patients: Implications for Partner Notification and Patient-Delivered Partner Therapy. *Archives of Sexual Behavior* 47 (2): 481–492.

Jo, Eun A. 2020. South Korea's Experiment in Pandemic Surveillance. *The Diplomat* [online] (April 13). https://thediplomat.com/2020/04/south-koreas-experiment-in-pandemic-surveillance/. Accessed 15 July 2020.

Kampf, Antje. 2008. A 'little world of your own': stigma, gender, and narratives of venereal disease contact tracing. Health: An Interdisciplinary Journal for the Social Study of Health. *Illness and Medicine* 12 (2): 233–250.

Kazanjian, Powel. 2012. The AIDS Pandemic in Historic Perspective. *Journal of the History of Medicine and Allied Sciences* 69 (3): 351–382.

Kass, Nancy, and Andrea Carlson Gielen. 1998. The Ethics of Contact Tracing Programs and Their Implications for Women. *Duke Journal of Gender Law & Policy* 5: 89–102.

Kegeles, Susan M., Joseph A. Catania, Thomas J. Coates, Lance M. Pollack, and Bernard Lo. 1990. Many people who seek anonymous HIV-antibody testing would avoid it under other circumstances. *AIDS* 4 (6): 585–588.

Keane, Mary Beth. 2013. *Fever*. New York: Scribner's.

Kim, Max S. 2020. Contact Tracing. *The New Yorker* [online] (April 17). https://www.newyorker.com/news/news-desk/seouls-radical-experiment-in-digital-contact-tracing.

Lambert, Bruce. 1989. AIDS Insurance Coverage is Increasingly Hard to Get. *The New York Times* [archive online], https://www.nytimes.com/1989/08/07/us/aids-insurance-coverage-is-increasingly-hard-to-get.html. Accessed 15 July 2020.

Leavitt, Judith Walzer. 1997. *Typhoid Mary: Captive to the Public's Health*. Boston: Beacon Press.

Leonard, Thomas C. 2016. *Illiberal Reformers: Race, Eugenics & American Economics in the Progressive Era*. Princeton: Princeton University Press.

Levine, Carol, and Ronald Bayer. 1989. The Ethics of Screening for Early Intervention in HIV Disease. *American Journal of Public Health* 79 (12): 1661–1667.

Levine, Philippa. 1996. Rereading the 1890s: Venereal Disease as "Constitutional Crisis" in Britain and British India. *The Journal of Asian Studies* 55 (3): 585–612.

Lin, Laura, and Bryan A. Liang. 2005. HIV and Health Law: Striking the Balance Between Legal Mandates and Medical Ethics. *Virtual Mentor* 7 (10): 687–692.

Lin, Liza, and Timothy W. Martin. 2020. How Coronavirus is Eroding Privacy. *The Wall Street Journal* [online] (April 15). https://www.wsj.com/articles/coronavirus-paves-way-for-new-age-of-digital-surveillance-11586963028. Accessed 15 July 2020.

Lindsay, Matthew J. 1998. Reproducing a Fit Citizenry: Dependency, Eugenics, and the Law of Marriage in the United States, 1860–1920. *Law and Social Inquiry* 23: 541–585.

Luker, Kristin. 1998. Sex, Social Hygiene, and the State: The Double-Edged Sword of Social Reform. *Theory and Society* 27 (5): 601–634.

Lurie, Peter, and Sidney M. Wolfe. 1997. Unethical Trials of Interventions to Reduce Perinatal Transmission of the Human Immunodeficiency Virus in Developing Countries. *New England Journal of Medicine* 337: 853–856.

Magaziner, Sarah, Madeline C. Montgomery, Thomas Bertrand, Daniel Daltry, Heidi Jenkins, Brenda Kendall, Lauren Molotnikov, Lindsay Pierce, Emer Smith, Lynn Sosa, Jacob J. van den Berg, Theodore Marak, Don Operario, and Philip A. Chan. 2018. Public Health Opportunities and Challenges in the Provision of Partner Notification Services: The New England Experience. *BMC Health Services Research* 18: 75. https://doi.org/10.1186/s12913-018-2890-7. Accessed 15 July 2020.

Mann, Jonathan. 1987. AIDS—A Global Challenge. *Health Education Journal* 46 (2): 43–45.

Marineli, Filio, Gregory Tsoucalas, Marianna Karamanou, and George Androutsos. 2013. Mary Mallon (1869–1938) and the History of Typhoid Fever. *Annals of Gastroenterology* 26 (2): 132–134.

McGann v. H & H Music Company, 946 F.2d 401 (5th Cir. 1991).

McNeil, Donald G., Jr. 2018. He Took a Drug to Prevent AIDS. Then He Couldn't Get Disability Insurance. *The New York Times* (Feb. 12) [online]. https://www.nytimes.com/2018/02/12/health/truvada-hiv-insurance.html. Accessed 15 July 2020.

Molina, Jean-Michel, Isabelle Charreau, Bruno Spire, Laurent Cotte, Julie Chas, Catherine Capitant, Cecile Tremblay, Daniela Rojas-Castro, Eric Cua, Armelle Pasquet, Camille Bernaud, Claire Pintado, Constance Delaugerre, Luis Sagaon-Teyssier, Soizic Le Mestre, Christian Chidliac, Gilles Pialoux, Diane Ponscarme, Julien Fonsart, David Thompson, Mark A. Wainberg, Veronique Doré, and Laurence Meyer. 2017. *The Lancet: HIV* 4 (9): PE402–PE410.

Moore, Kristine, Jill DeBoer, Richard Hoffman, Patrick McConnon, Dale Morse, and Michael Osterholm. 2020. *COVID-19: The CIDRAP Viewpoint, Part 4: Contact Tracing for COVID-19: Assessing Needs, Using a Tailored Approach* (June 2). CIDRAP. https://www.cidrap.umn.edu/sites/default/files/public/downloads/cidrap-covid19-viewpoint-part4.pdf. Accessed 14 July 2020.

Mozur, Paul, Raymond Zhong and Aaron Krolik. 2020. In Coronavirus Fight, China Gives Citizens a Color Code, With Red Flags. *The New York Times* [online] (March 1). https://www.nytimes.com/2020/03/01/business/china-coronavirus-surveillance.html. Accessed 15 July 2020.

National Cancer Institute (NCI). 2019. About the SEER Program. https://seer.cancer.gov/about/. Accessed 15 July 2020.

Nield, David. 2020. How Covid-19 Contact Tracing Works on Your Phone. *WIRED* [online] (June 7). https://www.wired.com/story/covid-19-contact-tracing-apple-google/. Accessed 15 July 2020.

North, Richard L., and Karen J. Rothenberg. 1993. Partner Notification and the Risk of Domestic Violence against Women with HIV Infection. *New England Journal of Medicine* 329: 1194–1196.

Nurse-Findlay, Stephen, Melanie M. Taylor, Margaret Savage, Maeve B. Mello, Sanni Saliyou, Manuel Lavayen, Frederic Seghers, Michael L. Campbell, Françoise Birgirimana, Leopold Ouedraogo, Morkor Newman Owiredu, Nancy Kidula, and Pyne-Mercier Lee. 2017. Shortages of Benzathine Penicillin for Prevention of Mother-to-Child Transmission of Syphilis: An Evaluation from Multi-Country Surveys and Stakeholder Interviews. *PLoS Medicine* 14 (12): e1002473. https://doi.org/10.1371/journal. Accessed 15 July 2020.

Otterman, Sharon. 2020. N.Y.C. Hired 3,000 Workers for Contact Tracing. It's Off to a Slow Start. *The New York Times* [online] (June 21). https://www.nytimes.com/2020/06/21/nyregion/nyc-contact-tracing.html. Accessed 15 July 2020.

Padgug, Robert A., Gerald M. Oppenheimer, and Jon Eisenhandler. 1993. AIDS and Private Health Insurance: A Crisis of Risk Sharing. *Cornell Journal of Law and Public Policy* 3 (1): 55–81.

Parker, Richard. 2002. The Global HIV/AIDS Pandemic, Structural Inequalities, and the Politics of International Health. *American Journal of Public Health* 92 (3): 343–347.

Parker, Richard, and Peter Aggleton. 2003. HIV and AIDS-related stigma and discrimination: a conceptual framework and implications for action. *Social Science & Medicine* 57 (1): 13–24.

Pérez-Stable, Eliseo J. 1991. Cuba's Response to the HIV Epidemic. *American Journal of Public Health* 81 (5): 563–567.

Perry, Nicole. 2015. Diseased Bodies and Ruined Reputations: Venereal Disease and the Construction of Women's Respectability in early 20th Century Kansas. https://kuscholarworks.ku.edu/bitstream/handle/1808/25635/Perry_ku_0099D_14388_DATA_1.pdf?sequence=1. Accessed 15 July 2020.

Poirier, Suzanne. 1995. *Chicago's War on Syphilis, 1937–1940: The Times, the Trib, and the Clap Doctor*. Urbana/Chicago: University of Illinois Press.

Reuters. 2020. Latvia to Launch Google-Apple Friendly Coronavirus Contact Tracing App. Reuters Technology News [online] (May 25). https://uk.reuters.com/article/us-health-coronavirus-tech-latvia/latvia-to-launch-google-apple-friendly-coronavirus-contact-tracing-app-idUKKBN23118I. Accessed 15 July 2020.

Rodgers, Bethany. 2020. Utah's Expensive Coronavirus App Won't Track People's Movements Anymore, Its Key Feature. Salt Lake Tribune [online] (July 11, 2020). https://www.sltrib.com/news/politics/2020/07/11/states-m-healthy-together/. Accessed 15 July 2020.

Rothenberg, Karen H., and S.J. Paskey. 1995. The Risk of Domestic Violence and Women with HIV Infection: Implications for Partner Notification, Public Policy, and the Law. *American Journal of Public Health* 85 (11): 1569–1576.

Rothschild, Bruce M. 2005. History of Syphilis. *Clinical Infectious Disease* 40 (10): 1454–1463.

Samuels, Michael E., Jonathan Mann, and C. Everett Koop. 1988. Containing the Spread of HIV Infection: A World Health Priority. *Public Health Reports* 103 (3): 221–223.

Schwartzbaum, Judith A., John R. Wheat, and Robert W. Norton. 1990. Physician Breach of Patient Confidentiality among Individuals with Human Immunodeficiency Virus (HIV) Infection: Patterns of Decision. *American Journal of Public Health* 80 (7): 829–834.

Scitovsky, Anne A. 1988. The Economic Impact of AIDS in the United States. *Health Affairs* 7 (4): 32–45.

Shaffer, D.N., V.N. Yebei, J.B. Ballidawa, John E. Sidle, J.Y. Greene, Eric M. Meslin, Sylvester J.N. Kimaiyo, and William M. Tierney. 2006. Equitable Treatment for HIV/AIDS Clinical Trial Participants: A Focus Group Study of Patients, Clinician Researchers, and Administrators in Western Kenya. *Journal of Medical Ethics* 32: 55–60.

Shrank, Alan B. 1987. Letter: Is Testing for HIV Without Consent Ever Warranted? *British Medical Journal* 294 (14 Feb.): 445.

Singer, Natasha. 2020. Google Promises Privacy With Virus App but Can Still Collect Location Data. *The New York Times* [online] (July 20). https://www.nytimes.com/2020/07/20/technology/google-covid-tracker-app.html. Accessed 20 July 2020.

Smith, David. 2011. Pfizer Pays Out to Nigerian Families of Meningitis Drug Trial Victims. *The Guardian* [online] (Aug. 12). https://www.theguardian.com/world/2011/aug/11/pfizer-nigeria-meningitis-drug-compensation. Accessed 15 July 2020.

Snyder, Lynne Page. 1995. The Career of Surgeon General Thomas J. Parran, Jr., MD, (1892–1968). *Public Health Reports* 110: 630–632.

Soper, George A. 1939. The Curious Career of Typhoid Mary. *Bulletin of the New York Academy of Medicine* 15 (10): 698–712.

Southern Illinoisan v. Illinois Department of Public Health, 844 N.E. 2d 1 (Ill. 2006).

Steven Joffe, Holly Fernandez Lynch. 2018. Federal Right-to-Try Legislation — Threatening the FDA's Public Health Mission. *New England Journal of Medicine* 378 (8): 695–697.

Swartz, Martha. 1987. AIDS Testing and Informed Consent. *Journal of Health Politics, Policy and Law* 13 (4): 607–621.

Symanski, Richard. 1974. Prostitution in Nevada. *Annals of the Association of American Geographers* 64 (3): 357–377.

Tarasoff v. Regents of the University of California, 551 P.2d 334 (1976).

Taylor, Josh. 2020. How Did the Covidsafe App Go from Being Vital to Almost Irrelevant? *The Guardian* [online] (May 23). https://www.theguardian.com/world/2020/may/24/how-did-the-covidsafe-app-go-from-being-vital-to-almost-irrelevant. Accessed 15 July 2020.

Tomes, Nancy. 1997. Moralizing the Microbe: The Germ Theory and the Moarl Construction of Behavior in the Late-Nineteenth-Century Antituberculosis Movement. In *Morality and Health*, ed. Allan M. Brandt and Paul Rozin, 271–296. New York: Routledge.

Tramont, Edmund C., and Shant S. Bohajian. 2010. Learning from History: What the Public Health Response to Syphilis Teaches Us About HIV/AIDS. *Journal of Contemporary Health Law & Policy* 26: 253–299.

UNAIDS. 2018. ABOUT: Saving Lives, Leaving No One Behind. http://www.unaids.org/en/whoweare/about. Accessed 15 July 2020.

———. 2018a. Miles To Go: Closing Gaps, Breaking Barriers, Righting Injustices. http://www.unaids.org/sites/default/files/media_asset/miles-to-go_en.pdf. Accessed 15 July 2020.

Varmus, Harold, and David Satcher. 1997. Ethical Complexities of Conducting Research in Developing Countries. *New England Journal of Medicine* 337: 1103–1005.

Vreeman, Rachel, Eunice Kamaara, Allan Kamands, David Ayuku, Winstone Nyandiko, Lukoye Atwoli, Samuel Ayaya, Peter Gisore, Michael Scanlon, and Paula Braitstein. 2012. A Qualitative Study Using Traditional Community Assemblies to Investigate Community Perspectives on Informed Consent and Research Participation in Western Kenya. *British Medical Journal Medical Ethics* 13: 23.

Winston, Morton, and Sheldon Landesman. 1987. Case Studies: AIDS and the Duty to Protect. *Hastings Center Report* 17 (1): 22–23.

World Medical Association. 2019. Declaration of Helsinki. https://www.wma.net/policies-post/wma-declaration-of-helsinki-ethical-principles-for-medical-research-involving-human-subjects/. Accessed 15 July 2020.

Xinhuanet. 2020. Alipay Health Code Landed in 100 Cities in 7 Days, Digital Epidemic Prevention Ran Out of "Chinese Speed" (Feb. 19). http://www.xinhuanet.com/tech/2020-02/19/c_1125596647.htm.

Yang, Y. Tony, and Kristen Underhill. 2018. Rethinking Criminalization of HIV Exposure—Lessons from California's New Legislation. *New England Journal of Medicine* 378: 1174–1175.

Zhong, Raymond. 2020. China's Virus Apps May Outlast the Outbreak, Stirring Privacy Fears. *The New York Times* [online] (May 26). https://www.nytimes.com/2020/05/26/technology/china-coronavirus-surveillance.html.

Chapter 4
Surveillance and Equity: Identifying Hazards in the Environment

In the preceding chapter, we emphasized that surveillance must not result in people being harmed because of who they are or the groups with which they are associated. In this chapter, we take up another injustice in surveillance: where surveillance looks. Surveillance may be directed towards protecting the health of some, but not the health of others in the community. These inequities in surveillance may undermine trust, particularly if data gathered from those who are not benefited by surveillance is used to benefit others who are better off.

4.1 Health Equity

Health equity is a core goal of public health practice. The WHO Guidelines place improving equity—that is, reducing inequality that is morally relevant—as a central goal of surveillance:

> Public health surveillance can further the pursuit of equity by identifying the particular problems of disadvantaged populations, including global communities, providing the evidence for focused health campaigns and identifying the basis of unfair differences in health. (WHO 2017a, p. 21)

On this view, what makes inequality morally relevant is its unfair impact on disadvantaged populations.

The Robert Wood Johnson Foundation (RWJ) discussion of wealth and health inequity similarly emphasizes unfairness to disadvantaged populations:

> Health equity means that everyone has a fair and just opportunity to be as healthy as possible. This requires removing obstacles to health such as poverty, discrimination, and their consequences, including powerlessness and lack of access to good jobs with fair pay; quality education and housing; safe environments; and health care. For the purposes of measurement, health equity means reducing and ultimately eliminating disparities in health and its

determinants that adversely affect excluded or marginalized groups. (Braveman et al. 2018, p. 1; see also Braveman et al. 2017)

RWJ describes "excluded or marginalized groups" as those who have been "pushed to society's margins, with inadequate access to key opportunities"; they include people of color, people living in poverty especially across generations, religious minorities, people with physical or mental disabilities, LGBTQ persons, and women. (Braveman et al. 2017, p. 4) According to RWJ, in choosing where to focus, public health organizations may consider whether groups suffer multiple disadvantages and where maximal impact can be achieved, among other factors. (Braveman et al. 2017, p. 4) Also according to RWJ, public health should engage these groups in making decisions about surveillance and subsequent actions to improve health equity; we discuss the importance of engagement more fully in Chapter 7 on communities and consent.

Scholars of health equity such as Sir Michael Marmot emphasize that achieving equity requires far more than improving access to health care or improving health care quality. (Institute of Health Equity 2010; WHO Commission 2008). Beyond health care, social determinants of health such as education, employment, sustainable communities, and systemic racism, play major roles in health outcomes. Put succinctly, "health inequalities that could be avoided by reasonable means are unfair." (Institute of Health Equity 2010, 3).

Surveillance may help in improving health equity. Or, it may compound inequity, both for the information it reveals and the information it does not reveal. If population subgroups are left out of surveillance, information about their disadvantages may not come to light. Predictive algorithms applied to skewed data sets will yield skewed predictions; for example, concerns have been raised that the use of artificial intelligence algorithms in diagnoses of skin cancers will be less accurate among Blacks if they are developed using data primarily from Caucasian patients (Adamson and Smith 2018). If information about other population groups reveals disadvantage, they may be the recipients of efforts to improve equity—efforts that are well intentioned but that have the potential to further deepen the disadvantages of those who are never noticed. How information gained from surveillance is communicated poses an additional set of problems, among them concerns that it will be in languages or forms not well understood or risks that people who are already disadvantaged will be further targeted because of the information revealed.

Providing the information needed to assess inequity—how great it is, whether it is located in particular pockets of a population, and whether progress has been made in improving equity—is a core task of public health surveillance. With the emphasis on social determinants of health, surveillance may need to be broadened beyond health care and health status. Instead, circumstances of daily life and the environments in which people live will loom in importance. For this reason, we use surveillance of environmental hazards to health as our primary illustration for the discussion of equity in surveillance; readers should keep in mind, however, that similar issues will attend equity in surveillance of other topics such as contagious or chronic diseases.

4.2 Environmental Hazards and Public Health Surveillance

Along with preventing the spread of contagious disease, reducing threats from environmental hazards has been the other primary goal of traditional public health. These hazards may be found in the natural world or in the built environment and may be exacerbated by the interface between the two. Today, pressing needs to assess environmental hazards are complicated by impacts of climate change that are generating rapidly shifting and unanticipated health challenges and patterns of disease.

Identification of natural reservoirs of disease, such as swamps where mosquitoes breed or caves where bats dwell can be crucial information for averting the transmission of diseases to humans. Human invasion into previously unsettled areas can lead to the transmission of novel zoonotic infections such as Ebola from non-human animals to humans. Environmental toxins such as lead, mercury, or asbestos, also can prove devastating to human health. Microbes in soil or water can cause diseases such as tetanus.

Sanitary movements emerged across the centuries to address the hazards created by crowded living conditions, such as sewage, garbage, odors, and smoke. One of the most famous examples of public health success was Sir John Snow's isolation of the Broad Street pump as the source of a cholera epidemic in London in 1854 discussed in detail in Chapter 2; his removal of the pump handle stopped the epidemic from further spread. Sanitation was the goal most identified with improvements in living conditions in increasingly urbanized and crowded environments.

Today, many of the environmental concerns of the early sanitarians—sunlight, clean water, trash collection, and sewage disposal—would be recognizable as aspects of the social determinants of health. Surveillance that attends to health-affecting conditions in the natural or constructed environment may identify social inequities. Conversely, the failure to surveil for such conditions may also be inequitable.

Several features of environmental surveillance should be noted at the outset as relevant from the perspective of equity. Environmental surveillance may reveal information about benefits or burdens that cannot be confined to particular individuals or groups or that may be more particularized. Environmental surveillance may reveal information about hazards that are relatively confined geographically or that cannot be confined within political borders. And, environmental surveillance may press questions about the scope of public health and surveillance. As environmental surveillance observes the conditions in which people live, it may also be perceived as unduly intrusive or threatening.

4.2.1   Public Goods

An important complication of environmental surveillance is that the information it reveals may be about what are known as "public goods." Public goods are defined by economists as goods which by their nature cannot be restricted to individuals who pay for them. They are non-excludable—if they are provided for some they will also be enjoyed by others—and non-rivalrous—if some enjoy them others will still be able to enjoy them too. Clean air is a primary example. Air is ambient and everyone breathes it. The benefits of reductions in particulate emissions then cannot be confined to those who pay for them. Nor does one person's breathing clean air reduce the opportunity for others of doing so. Masks contrast: society could address the problem of air pollution by selling masks through the market to those who are able and willing to pay for them and who can deal with the burdens of wearing them. Presumably such market arrangements would also feature a variety of masks of different quality, so those who could pay more might have improved filtration capacity or masks of lighter weight or greater beauty.

Much of what may be done to address environmental issues such as clean air involves public goods in this sense. Everyone benefits—or loses—to at least some extent when information is gleaned about pollution levels that then leads to either increased enforcement of existing standards or imposition of heightened standards. So-called "free riders" can receive the benefit of air quality improvement without paying any of the costs of its production. This raises equity problems: some members of the population will bear the costs of cleaner air while others will enjoy it for free, and both costs and benefits of the good's production may also be inequitably enjoyed.

Transportation policy is a good example of how the costs of producing clean air may not be distributed equitably. Suppose surveillance reveals high levels of the fine particulate emissions associated with automobile exhaust. Policies that rely on voluntary behavior change such as encouraging the use of public transit or limiting driving trips will allow some to free ride on the good efforts of others. Other policy options may seek to spread costs more widely, such as a gas tax that increases the costs of driving private vehicles, but may still have inequitable effects. Those who are better off in the population may be able to pay the surcharge to drive private vehicles whereas those with fewer resources may find themselves consigned to using public transit. Although the wealthier in this scenario do pay increased taxes, the impact of these costs on them may be far more limited than the full costs of shifting to public transit for those who are worse off and who may experience infection risks on crowded buses, longer commutes to jobs, reduced employment opportunities because public transit is not available to desirable jobs, or difficulties in finding child care that is consistent with transit schedules. Redistributive adjustments such as directing the increased gas tax revenues to improvements in public transit may mitigate the inequities, but these adjustments may not be politically practical, may not solve problems such as infection risks, and in any event may not fully make up for the increased burdens.

Also importantly from the perspective of equity, the benefits and burdens of these improvements may not be distributed equally. For example, emission reductions may not create the same levels of pollution reduction in all parts of an urban area; parts of the metropolis located at higher altitudes may benefit more than lower lying areas where remaining pollution settles. Unfortunately, these distributive differences may bear more heavily on those who are already worse off: people who can only afford less expensive property near toxic sites, low-lying marshland, or areas of sewage disposal. To take another example, rising ocean levels due to global climate change present especially sharp challenges to the availability of clean water in impoverished areas of the globe where droughts are severe and flooding contaminates water sources. (UN Water 2019).

4.2.2 Political Borders

Some factors that affect disease prevalence or severity are relatively localized within political borders but others are not. Once again, air quality is a good illustration: local conditions may affect the extent to which particulates are trapped by inversions. Depending on their mass, particulates may remain locally or travel far and wide. Seeds, insects, microbes, and animals cross borders, too; plague-carrying fleas hopped political jurisdictions with impunity over the centuries. Rivers may run between political jurisdictions and effluents discharged upstream will make their way through downstream areas. If an upstream or upwind jurisdiction does not become aware of a toxic emission, it may fail to provide critical information to downstream or downwind jurisdictions in time to prevent serious harm to inhabitants of these lower regions. Even apparently local discharges such as mine tailings may leak into ground water or aquafers and eventually end up across state or national lines. Climate change may be the most universal of all such examples; emissions behaviors in high-consumption countries such as the United States will affect everyone regardless of the jurisdiction in which they live. These changes will have major impacts on population health, from heat waves to changes in the habitat of mosquitoes or other disease carrying insects.

Political borders present problems for the equity of surveillance in several ways. According to the WHO (2019), on average countries across the globe have achieved 71% of the surveillance attributes required by 2018. Africa region countries have achieved the lowest level of required capacities, at 59%. These capacities include financing, human resources, laboratory capabilities, and monitoring at the border. They also include the ability to detect zoonotic events and the possibility of animal-human transmission, unsafe food, chemical events, and radiation exposures. Importantly for others, they also include risk communication, which was at only 57% globally. This figure means that just over half of the countries across the globe had achieved WHO required attributes for risk communication; while the Africa region was the lowest at 39%, South-East Asia was at 60% and the Americas and the

Western Pacific were at 63%–less than two-thirds. Only Europe, at 65%, approached two-thirds of countries achieving the requisite abilities.

Many different factors account for these gaps. Some areas are simply more difficult to surveille because of the nature of the terrain or the lack of infrastructure such as roads or communication systems. Cultural factors or mistrust may explain why people are unwilling to share information. Jurisdictions may have very different resources available to devote to surveillance and limited health care infrastructure to serve as points of contact for information. Jurisdictions also may make different judgments about the resources to devote to surveillance, what to surveille, or whether to reveal the results of surveillance. The current International Health Regulations encourage States Parties to collaborate in detection and to mobilize resources for developing countries to build and maintain their surveillance capabilities (WHO 2016, Article 44). However, this provision is only an encouragement and does not impose requirements on States Parties. The upshot for equity is that the costs of developing surveillance capabilities may be more burdensome in comparison to resources in some countries than in others.

On the other side, failures to surveille may also have inequitable consequences. Countries with greater resources may be less vulnerable to detection failures elsewhere as their better detection capabilities may help to fill in the gaps. But this will not be perfect, especially for conditions that require detection in real time to prevent spread; it may be even more deadly for countries lacking their own surveillance capabilities. The spread of the Ebola epidemic from Guinea across West Africa in 2014 was one recent illustration of the tragic potential of such failures. The spread of COVID-19 is an even greater tragedy, although perhaps attributable more to delays in communicating information and acting on it, rather than to an initial failure to recognize the emerging infection. A further complication for inequity is that countries with more limited surveillance capabilities may also be more heavily impacted by factors that give rise to the need for surveillance, such as swampland, downstream territory, or a warm climate.

We would be remiss, however, by failing to note the potential benefits of jurisdictional differences for surveillance. In the United States, it has become canonical to assert that federalism allows for the states to function as laboratories. States may experiment with different surveillance methods and learn from one another. Jurisdictions may also have the local knowledge that enables them to take local conditions into account, from the landscape to cultures and religions. To the extent that it is easier to develop trust in local governmental units than in units that function from afar, surveillance may be better carried out at local rather than national or global levels.

4.2.3 Intrusion

At first appearance, environmental surveillance may seem to pose far less in the way of privacy challenges than forms of surveillance that require contact with individuals or information from their health records. Monitors of air or water quality set up in neighborhoods, for example, do not collect information about the local inhabitants. Yet these indicators may be indirect sources of information about people who live nearby that may result in deleterious consequences for them. How this information is communicated may be important, too. Consider as an example people evacuated from the area around the Chernobyl nuclear power plant disaster. If people are labeled "Chernobyl survivors" without further explanation or understanding, the result could be fear or stigmatization. Fear could result, for example, by confusion between having suffered a radiation exposure and being radioactive; this confusion is widespread, as illustrated also by suspicions of food that has been sanitized by radiation exposure as subjecting those who eat it to radiation. Labelling COVID-19 the "Wuhan virus" or the "Chinese virus" has been used to stigmatize, too.

Environmental surveillance may also pose intrusion risks if information gleaned from the environment suggests the need to inspect homes. Surveillance of water discharged into a sewer from a home, for example, could indicate the presence of disease in the home and lead to efforts to search the home. Technologies for remote detection of chemical and biological agents are developing rapidly and may yield information that suggest further exploration of intimate space.

Information gleaned from environmental surveillance may ultimately suggest the need to access human bodies more intimately. Understanding what is called the "microbiome"—the thousands of different species of microorganisms that inhabit the human body—is just beginning but already has yielded important information about human health and disease. Each human has a unique microbiome, shaped by exposures, diet, and interactions with the person's genetic makeup. This microbiome is a delicate balance of deleterious and beneficial organisms that are critical to digestion and immune status among other aspects of human health. Research is yielding findings such as the relationship between the microbiome and susceptibility to diabetes or risks of preterm birth (iHMP Consortium 2019). Studies of the microbiome are also yielding information that may prove critical to addressing health inequities such as the increased prevalence of preterm birth among women of African ancestry. Shifts in the microbiome may be harbingers of shifts in human disease susceptibility and thus may become information critical to surveillance. Yet detection of these shifts may require samples of excrement or even samples obtained directly from areas of the human body such as the gut or the reproductive tract, both areas of the body regarded as particularly sensitive by many.

To summarize briefly: environmental surveillance is a critical aspect of public health surveillance. It may bring to light health inequities associated with social determinants of health. Yet it may itself be inequitable in both the burdens and benefits it imposes. These inequities are complicated by the fact that some environmental goods such as clean air are goods that can be experienced by everyone, albeit to different extents, regardless of whether they contribute to production of the goods. They are also complicated by the fact that environmental factors often cross political jurisdictions which may have very different capacities and approaches to surveillance. Finally, although environmental surveillance may on the surface not appear to raise privacy questions, because the information it gathers in the first instance may not be about individuals at all, it would be misleading to conclude that privacy is not implicated by this form of surveillance. Environmental surveillance may lead indirectly to conclusions about people and the need for further examination of their living spaces or even their bodies. We now turn to deeper exploration of equity in surveillance, using the example of how water is, or is not, surveilled.

4.3 Water Surveillance Disparities

Among the most serious problems for disadvantaged populations is the lack of clean water, a basic necessity (WHO Commission 2008, p. 14). The UN Sustainable Development Goal for clean water and sanitation targets achieving access to safe and affordable drinking water, reducing polluting and minimizing release of hazardous chemicals, and achieving access to adequate and equitable sanitation. (UN 2018a) Across the globe today, three in ten people lack access to safe drinking water, six in ten lack access to safe sanitation, 2.4 billion people lack even basic sanitation, and nearly 1000 children die daily from preventable water-related diarrheal diseases. Safe water is not just a problem for poorer nations, however. In the United States, children in Flint, Michigan, were exposed to high levels of lead for over a year after a change in the source of their water. A 2018 report of the European Environmental Agency rated only 38% of Europe's surface waters as in good chemical status (EEA 2018, p. 6).

Environmental surveillance is critical to identify health hazards such as unsafe water. Threshold problems of equity attend how this surveillance is deployed, as environmental hazards for some may pass without attention while hazards for others are scrupulously monitored. Further problems of equity are consequent to what the information brought by surveillance reveals, such as precipitous declines in property values after information comes to light that a development was built in an area of high toxicity. Equity issues also arise with respect to possible responses to the hazards identified through environmental surveillance. The account of surveillance equity we develop here is rooted in the understanding that equity may be constructed differently in contexts of injustice rather than for a more perfect world. Public health surveillance today takes place in a world of deep poverty and stunning affluence; it will prove unsustainable if it fails to take these inequities into account.

Many factors attend gaps in surveillance equity. Water systems may be fragmented and privately rather than publicly run. Aging infrastructures are expensive to maintain and bad news about their condition may not be politically welcomed. Water is an economical and often unobserved site for waste disposal from profitable industries such as mineral extraction or large-scale agriculture. Water is used in fracking or other methods of petrochemical extraction. Drinking water is especially difficult to surveil as it may come from multiple small sources: rainfall collection, wells, streams and rivers, small community water systems, and other dispersed sources. Yet, as we explore below, serious gaps in water surveillance are an illustrative example of problems of equity in surveillance.

4.3.1   Clean Water, the UN, and the WHO

In 2010, the United Nations General Assembly adopted Resolution 64/292 recognizing safe and clean drinking water and sanitation as a human right that is essential for the full enjoyment of life and all human rights (United Nations General Assembly 2010). Advocates of water justice pushed for the resolution. Support for the resolution was not unanimous, however. Most notably, both the United States and Canada abstained from the vote. As indicated above, a goal of the United Nations Sustainable Development Goals is access to clean water and sanitation (UN 2018b). The 2018 report on progress towards this goal indicated the need for accelerated efforts if these goals are to be met and noted especially the damaging effects of conflict and instability. It also singled out the need for improved management within countries of water resources, accelerated cooperation among countries sharing water resources, and a disturbing decline of more than 25% of official development assistance for water-related activities (United Nations 2018b).

Beginning with 1990, the WHO and UNICEF established a water supply monitoring program, publishing regular updates during the Millennium Development Goal period. These updates are continuing with the 2030 Agenda for Sustainable Development, which also emphasizes the importance of safe water. However, despite these efforts, reliable estimates for access to safely managed drinking water were available for only 96 countries (35% of the global population) (WHO 2017b, p. 3). While these figures suffice to indicate broad problem areas, they are by no means adequate; in the judgment of the WHO, "there are major data gaps, and effective monitoring of inequalities…will require significant improvements in the availability and quality of data underpinning national, regional and global estimates of progress" (WHO 2017b, p. 9). Data are often not subdivided by income or by subregion even when national level data are available; disparities between urban and rural areas are thus comparatively under-recognized (WHO 2017b, p. 34). Data gaps are also particularly noteworthy for schools and health care facilities, institutional types where access to safe water, sanitation, and hygiene facilities are critical because of their impact on children and people who are already frail (WHO 2017b, p. 44).

These gaps in knowledge cannot be attributed solely to disparities between the developed and the developing world. In the U.S., the Safe Drinking Water Act of 1974 requires monitoring and notifying consumers of safety violations with the potential to cause serious health effects. However, the requirement of public notice when drinking water exceeds lead action levels was only added by Congress after the devastating discovery of lead levels in the water system in Flint, Michigan (Tiemann 2017). The now-infamous example of Flint was not only a failure of water system management but also a failure to surveil drinking water conditions in a largely poor and minority community.

4.3.2   Flint, Michigan: A Surveillance Failure in a Wealthy Country

The serious effects of exposure to lead are well-known. In adults, they include abdominal pain, fatigue, headaches, irritability and aggression, loss of memory, pain in the extremities, nervous system damage, infertility and stillbirth. Children and developing fetuses are affected even more severely; risks of lead exposure include reduced cognitive function, mental retardation, inattention, hyperactivity, and antisocial behavior (CDC 2017).

Flint is a city of about 100,000 located in central Michigan, just over an hour's drive from Detroit. Although the site where General Motors was founded, Flint has become chronically depressed as automobile factories have closed and moved elsewhere. Today, Flint is a majority Black city, with a median household income under $30,000/year. Nearly 40% of Flint's residents live below the poverty level (US Census Bureau 2019).

In 2014, Flint's water supply was shifted from Detroit to the local Flint River to save costs. At the time, Flint had the highest cost to consumers of all water districts in the nation. Flint's water problems thus stand out in more than one way. After the shift, the corrosive river water leached lead from aging city pipes and bacterial contaminants entered the water supply. At one point, residents were urged to boil water because of bacterial contamination. Water in several homes tested at very high lead levels, yet city and state agencies continued to reassure city residents that their water was safe. After a water quality expert called attention to the role of corrosion and a group of local physicians reported high lead levels in children (Erb 2015), Flint residents were finally informed about the problems with their water and its likely cause (Hanna-Attisha et al. 2016). Fifteen state and local officials were criminally charged as a result of their actions to conceal the problems with the water (Egan 2017). A significant decline in reading scores among Flint children may be attributable to the lead exposure (Atkin 2018). So may be higher rates of stillbirths and miscarriage among Flint women (Grossman and Slusky 2017). The Michigan Civil Rights Commission (2017) has concluded that systemic racialization rather than intent by individual bad actors played a critical role in the water crisis. The city of Flint has a long history of racially segregated housing and tax structures that redirected resources from the city to the more affluent and white surrounding suburbs. These structures were largely responsible for the economic woes that had placed the city into the receivership arrangement that precipitated the water crisis (Highsmith 2015).

But Flint is not an isolated case. Many US cities with aging infrastructure have corroding pipes that allow lead and other contaminants to enter drinking water. Residents have learned about these risks belatedly or in some cases not at all (Wines and Schwartz 2016). Reports about schools, too, indicate high lead levels in drinking fountains and other water sources used by children in cities such as Jersey City, New Jersey (Wines et al. 2016).

At present, the U.S. Environmental Protection Agency (EPA) lead standard for safe drinking water is 15 parts per billion. Critics question this standard as based primarily on apparent existing levels in housing rather than evidence of safety (Wines and Schwartz 2016). Current regulatory structures leave out many municipal water systems from monitoring requirements, depending on the source of their water (Wines et al. 2016). Schools also are typically not water suppliers regulated by the U.S. Safe Drinking Water Act—exactly the kind of omission of concern to the WHO report singling out the gaps in knowledge about the safety of water supplies to hospitals or schools. Schools do come under the lead monitoring and reporting requirements of the U.S. Lead Contamination and Control Act, which include the requirement to replace drinking water fixtures that contain excessive lead amounts. However, monitoring in schools has been uneven at best (Wines et al. 2016). Replacing old lead pipes in cities or drinking fountains in schools is expensive. There is also significant political opposition to EPA involvement in setting local standards.

These reports of lead contamination in aging U.S. cities take place against a background of historical injustices, economic dislocation, and settlement patterns that have affected Blacks more severely than other groups. These urban areas also have aging infrastructures and high percentages of older schools. During the 1950s, they were sites for freeways built to improve transit for those living in more affluent suburbs; these freeways are an additional major source of lead exposure. Jersey City is but one case in point, with a majority of schools over 80 years old and a freeway extension cutting through it that has contaminated local parks used by children with lead (Mota 2016). Income inequality has concentrated poorer residents in areas of cities where housing costs are lowest. Redlining—the practice of rating neighbourhoods or borrowers in a manner that makes mortgage financing difficult or more expensive to obtain—functioned historically to concentrate Blacks in these areas. Not only a practice of private banks, redlining was an official practice of the Federal Housing Administration, which color-coded neighbourhoods according to their perceived risks to investors (Rothstein 2017). Once thought to have faded into history, discriminatory redlining appears to have been ongoing in areas of the country such as urban new Jersey, as a recent settlement with that state's largest savings bank reveals (Swarns 2015).

The problem of lead in drinking water has been known for many years, as have the background factors that have put poor and minority children at greater risk of significant exposure than their more fortunate counterparts. But the U.S. informational infrastructure about exposures to lead—and other contaminants as well—has seriously lagged. There is no systematic way of collecting information about lead levels in many water supplies, including those serving schools. Testing is required in only a small number of homes in large water systems, and then at quite extended intervals and based on protocols that are subject to criticism (Wines and Schwartz 2016). Baseline data to assess the impact of the Flint water crisis had to be extrapolated from a mixture of documents not entirely consistent in their methods of data collection (Drum 2016). Yet these data are fundamental to health, especially for children.

In the background of the structure of water surveillance in the U.S. is that water sanitation and delivery to customers is highly complicated and fragmented. The U.S. has some 52,000 water districts. Only about 400 of these serve more than 100,000 persons. Beginning in the 1920s during the Progressive Era, water supply, quality and treatment increasingly became publicly owned and managed. Progressives argued that safe drinking water needs were more likely to be addressed by the public sector than the private sector. Today, however, the number of privately-owned water districts is greater than the number of public districts in 17 states, Puerto Rico, and the District of Columbia. In many rural and suburban areas, small privately-run water districts are large in number but small in the number of households and business served. The number of private sector firms continues to contract across the country. Federal and to a lesser extent state laws govern water quality and supply. Water district managers recognize that maintaining both adequate supplies and reliable water quality requires increased funding but are reluctant to rely on increased fees for water usage to meet demand. They are seeking federal and state support, arguing at the federal level that water quality is a national responsibility given standards are federally determined.

Concerns over water quality have emerged across the U.S., particularly since the Flint crisis. Community water systems located in rural areas have the highest percentage of violations, over 9%. The southwest is the region of the country with the highest patterns of repeat violations. Minority communities are also more likely to experience water quality violations (Allaire et al. 2018).These disparities may reflect a failure of state level regulatory competence caused by underfunding, disinterest, or these districts' lesser ability than metropolitan water districts with their greater resources and political capability to secure federal resources. The funding ethics of federalism plays a role too, as federal authorities may establish regulatory regimes but anticipate that the state will significantly contribute to their realization. Or, states may lack sufficient domestic support or resources for the infrastructural investment to make to meet federal expectations.

4.4 Inequity in Safe Water Surveillance

These failures to surveille water quality are clearly harmful to human health, but it is a further question whether they are inequitable. Not every failure that is deleterious to health is also a wrongful inequality. For example, inadequate sanitation in an expensive private club may result in an outbreak of salmonella among its patrons but does not treat these patrons unfairly in comparison to others in the same society.

Nor are the club's patrons disadvantaged; indeed, it is their very privilege that exposes them to the private club's mistakes. Typhoid Mary infected her wealthy clients, but she did not subject them to inequity; indeed, it was only because of their privileged socioeconomic status that they were able to employ a private cook for their vacation homes in the first place.

In this section, we sketch three reasons for believing that failures of water surveillance are inequitable. First, they are unequally distributed among those who are better off and those who are worse off. Second, they have particularly serious consequences for those who cannot protect themselves, notably children. Finally, they are corrosive in the sense that they contribute to the worsening impacts of already existing inequities.

As used in the WHO Guidelines, "equity" refers to morally problematic inequality. Equity attends to inequalities that damage the most vulnerable unfairly. It can thus be situated as a particular issue within more over-arching theories of justice, without presupposing a full theory of justice. What is required for determining inequity is identification of those who are most disadvantaged and most vulnerable. This is not to say that more over-arching justice is irrelevant to inequity. Inequities take place in a wide variety of social circumstances and background justice or injustice is among these variables. Some inequities may be more or less isolated pockets within a more generally just society, as judged by one or more basic theories of justice. Other inequities—such as the ones identified in Flint—take place against a background of more extensive problems of social justice such as racism. As we described in the Introduction, the significance of such contexts of injustice is referred to in contemporary political philosophy as "non-ideal" theory or "partial compliance" theory, an approach that addresses the relevance of background injustice for addressing questions such as health inequity. We'll return in the final section of this chapter to the significance of framing water surveillance as a partial compliance problem after the fuller description of inequities in water surveillance that follows.

Several commentators have recently developed data-based accounts of water inequity both in the US and elsewhere. In so doing, these commentators have found the lack of available data methodologically challenging. Data are particularly lacking about some communities more likely to be disadvantaged—that is, there is an inequity in the available data. In a widely-referenced study, VanDerslice (2011) explored demographic characteristics of U.S. communities exposed to poor water quality, documenting their concentration in poor urban areas, rural areas with significant Latina/o populations especially in the San Joaquin Valley of California and colonias along the US-Mexico border, and tribal lands. Data needed for assessing demographic characteristics of communities served by community water systems, however, were not available so reliable disparity assessments were difficult to ascertain. Balazs and Ray (2014) use data from the San Joaquin Valley in California to explore how social and political factors affect the ability to address water contamination. Another recent study, building on VanDerslice, compared reported drinking water violations to population demographics. This study indicates that the failure in Flint may not be isolated; exposure to harmful contaminants as measured by repeat drinking water violations is significantly higher in community water systems serving poor and minority populations, especially when those populations have high percentages of people who lack health insurance (McDonald and Jones 2018). Another recent study of access to safe water in the U.S. underscores disparities in water affordability, especially in areas in need of infrastructure repair and with declining tax bases available to pay for these improvements; this study parallels what is known about problems with water affordability in developing areas of the world (Mack and Wrase 2017). These studies are suggestive of serious problems in water equity in the US and invite further, more systematic research.

A second reason why failures of adequate water surveillance are inequitable is their disproportionate impact on children. (WHO 2017a, p. 36) As highlighted by the example of Flint, children are especially susceptible to damage by high levels of lead that may be found in drinking water. Arsenic, another frequent contaminant of drinking water, is also associated with neurotoxicity to the developing brains of children, including deficits in intelligence and memory (Tolins et al. 2014). The behavioural and cognitive sequelae of these exposures can have lasting effects on the lifetime opportunities of these children. Children who have been damaged by lead exposures may be less able to succeed in school, as the initial data from Flint cited earlier in this chapter suggest, with compounding effects on employment and health status. Over a third of the million and a half deaths from diarrheal disease annually are in children under the age of 5; a significant portion of these illnesses can be prevented by safe water (WHO 2018). These waterborne illnesses also cause malnutrition and increased susceptibility to other illnesses, further reducing the opportunities of these children. The impact of shortages of safe water are also borne disproportionately by women who must use water in domestic labour and spend increasing amounts of time and energy bringing safe water to the household when it is available (Sultana 2018). The need for such domestic labour also has a disproportionate effect on girls, who are more likely to drop out of school to serve household needs.

Finally, through their widespread effect on opportunities, water disparities compound and deepen existing inequities. They are corrosive disadvantages in the sense outlined by Wolff and De-Shalit (2007): disadvantages that interact with other disadvantages such poverty or lack of education to further entrench the unjust circumstances of those who are subject to them. Given this potential for corrosive disadvantage, it is important to take care that surveillance does not add to the ways in which the situation of these already-vulnerable populations are compromised— as, for example, migrants might be targeted if disease transmission from unsafe drinking water is identified with them.

To summarize: for at least three reasons, disparities in access to safe drinking water and the failure to collect the information needed to identify these disparities violates basic equity. These reasons are that their burdens are distributed unequally, they disproportionately affect those who cannot protect themselves, and they compound already-existing injustice. To be sure, surveillance of water quality is difficult and expensive and may compete for other scarce funds available for surveillance. The point of this discussion has not been to resolve these difficult problems of equity within surveillance; suffice it to say for now that the health importance of access to safe water and the extent to which this access is itself an issue of equity creates a strong case for including water quality among an equitable system of surveillance.

4.5 Background Injustice and Surveillance Inequities

These inequities in access to safe water are in many cases attributable to existing injustice. Extractive industries create large quantities of waste and runoff that, if not properly managed, flood acids, mercury, and heavy metals into water supplies. Little, if any, of the benefits of these industries redound to the local population who are most affected by the water pollution (UN Environmental Programme 2017). Vast dams build for hydroelectric power are controversial not only because they displace local populations but also because they change patterns of erosion and create standing water that becomes new reservoirs for disease. In a thoughtful recognition of both the real benefits of dams for irrigation, water consumption, electricity generation, and flood control, dams have "been marred in many cases by significant environmental and social impacts which, when viewed from today's values, are unacceptable" (World Commission on Dams 2000, p. ix). To take just one disease-related effect of dams, increased rates of the parasitic disease schistosomiasis have been widespread after dam construction across Africa. Like diarrheal diseases, this condition causes higher morbidity among young children who are both more likely to exposed and who bear greater burdens of disease (Sokolow et al. 2017).

Occurring as they do within circumstances of injustice, these inequities present the question of the relevance of existing injustice to public policy, such as policy about surveillance. In the Introduction, we set out these two assumptions made by Rawlsian ideal theory (Rawls 1971): a compliance assumption and a favorable circumstances assumption. The compliance assumption hypothesized that sufficient numbers of relevant agents comply with demands of justice. The favorable circumstances assumption postulated that natural and historical conditions are favorable to the realization of justice.

Neither the compliance nor the favorable circumstances assumption holds in vast areas of the world today, in the assessment of a wide variety of theories of justice. Sketched in the briefest outline, the compliance assumption fails where corruption is widespread, where terrorist groups routinely threaten basic social institutions, or where corporations extract resources without any attention to the harms to health the resulting waste might cause. The favorable circumstances assumption fails where topsoil is so eroded that crops will not grow, where deforestation is extensive, where drought stretches on for years, or where lowlands are susceptible to widespread flooding after heavy rainfall or increasingly intense storm systems. The two failures intertwine, too, as when resource extraction leads to deforestation or crop failures augment the streams of refugee populations that are subject to exploitation by human traffickers.

As we sketched in the introduction, attention to non-ideal circumstances has been extensive in recent theorizing about justice. Several importantly different questions have drawn significant attention in these discussions. (Valentini 2012) Some consider what moral obligations individuals or societies have in circumstances of widespread failure of the compliance assumption, the favorable circumstances assumption, or both of these assumptions. For example, what obligations do others have to support surveillance in a country in which corruption is widespread and the results of surveillance may be inaccurate due to bribery, or where funding to support health improvements may be diverted? Should the primary goal of surveillance in such circumstances be only to identify hazards that could become health emergencies of international concern beyond that country's borders, as Ebola might become but rising rates of infant mortality would not? Or, should efforts still be directed to identifying hazards to the health of people within the country, such as improper sanitation that causes diarrheal disease in infants?

Other non-ideal theorists consider the extent to which feasibility of realization should constrain determinations about what should be done. Such feasibility considerations might include dangers to health workers conducting surveillance, difficulties in gathering needed data, or problems about whether any results of surveillance can be put into play to improve health outcomes. For example, what is the relevance to surveillance obligations of the lack of health infrastructure, of public reluctance to seek care, or of the likelihood of widespread efforts to conceal health information such as Ebola infection? One plausible response to these kinds of problems is that it would be preferable to spend scarce resources elsewhere, where benefits of surveillance are more likely to be gained. The result, however, might be further worsening of the health of already highly vulnerable populations. If so, equity might require efforts to address the circumstances that are creating obstacles to surveillance.

A still further aspect of non-ideal theory concerns the relationship between ideals and what ought to be done in non-ideal circumstances. Questions for surveillance in this regard would revolve around the relevance, if any, of ideal surveillance systems to the systems that should be put into place in the real world. Should ideal surveillance be the guide? What if there are side effects of guiding policy in the direction of the ideal, such as unanticipated risks to groups who are identified with particular health deficiencies? Or, should risks of further inequity to some constrain surveillance of others? On the views of some theorists, we should develop surveillance plans that strive to move us closer to ideal requirements. On the views of others, theorizing in non-ideal contexts is simply different from theorizing in ideal contexts and the latter cannot be a guide to the former (e.g. Wolff 2017; Sen 2011). In the remainder of this chapter, we develop examples of what equitable surveillance would require for each of these non-ideal theoretic questions. In so doing, we emphasize that non-ideal circumstances vary greatly; equity may require quite different approaches depending on the context.

4.6   Failures of Compliance: Water Surveillance or Health Emergencies of International Concern?

As suggested above, the compliance assumption fails in many countries today. This failure presents difficult questions about the obligations of those with resources to try to support surveillance where resources are necessary for it to take place. Surveillance is expensive and choices may be necessary about where to allocate scarce resources. So, it might be argued, surveillance of water conditions may assume lesser priority where corruption is likely to interrupt efforts at amelioration. Instead, surveillance should be directed only at conditions that are likely to spread and present dangers to others providing support for the surveillance. To put the point bluntly, surveillance should focus first on preventing pandemics such as COVID-19, especially in countries that do not cooperate and that depend on others for support of any public health efforts.

Perhaps most directly relevant to obligations to support water quality surveillance is the observation that in some countries monetary assistance aimed to improve water quality may be diverted by corrupt officials. The Berlin-based Water Integrity Network (WIN) monitors corruption as it affects water supplies. To quote just a few examples of such corruption documented in a 2016 WIN report:

- In Benin, € 4 million of Dutch funding vanished from the Ministry of Water in 2015.
- In Malawi, a reformed public financial management (PFM) system was misused to divert US$ 55 million from public funds to the private accounts of officials.
- The Nairobi City Water and Sewerage Company in Kenya loses 40 per cent of its supply to theft and leaks while poor residents are forced to buy water from vendors at ten to 25 times the price they would pay the water utility.
- In South Africa, eThekwini Metropolitan Municipality in KwaZulu-Natal lost more than a third of its water in one year because of illegal connections and vandalism, costing US$ 44 million. (WIN 2016, pp. 23–24).

Under such circumstances, are surveillance obligations changed by the presence of compliance failures? Some might argue that they are: that if funds to address water quality issues will be diverted, then surveillance of water quality in that area should be of lower priority than surveillance for other purposes, most importantly to identify public health emergencies that may become of international concern like a COVID-19 pandemic. The argument would be that resources for surveillance are scarce and choices must be made about where to spend them. If an area is likely to mis-spend any resources that would be provided to ameliorate bad water conditions, then perhaps surveillance of local conditions, such as water that exposes the local population to lead poisoning or infant diarrhea, should be of lower priority. Instead, resources should be spent elsewhere, where they could accomplish greater good. Moreover, if the resources are to be provided by countries that have interests in their own protection, perhaps those countries might justifiably prioritize any surveillance in a corrupt area that protects against risks of disease spread elsewhere. The high surveillance priority for identifying emerging global health emergencies of international concern would remain, but water quality issues affecting only the local population would be de-prioritized, on this view. The reply to this reasoning given by the WIN report is the importance of transparency: if we never know what is happening to the local population because of unsafe water, inequity will go unrecognized and unaddressed.

Documents from the WHO take differing positions on what is recommended in such circumstances of non-compliance. On the one hand, the WHO Guidelines judge that the global community has an obligation to support countries that lack adequate resources for surveillance in developing their surveillance capacities. Its reasoning in support of this view seems to extend beyond the interests of countries providing support in protecting themselves from disease spread. The Guidelines' stated rationale is that "equity provides the ethical foundations for claims to international support." (WHO 2017a, p. 32) If so, equity would provide an argument for supporting surveillance at least to identify the deficiencies and their effects on the health of the local population, even if nothing can immediately be done to improve the situation. This is the WIN position about the importance of transparency. The Guidelines also make clear that the priorities of the supported country may result in a different balance, although tradeoffs are not acceptable in situations of "gross injustice of violations of human rights" (WHO 2017a, p. 23). The Guidelines find further support for such surveillance both in preventing global spread of disease and in the need for international support to address environmental factors. (WHO 2017a, p. 25) Thus, the WHO Guidelines do not locate the obligation for support solely in whether there are risks of disease spread.

On the other hand, the WHO Guidelines also bring out considerations that would count against support for water issues affecting the local population in areas where corruption is widespread. The Guidelines state that when countries are failing to protect fundamental rights or interests, international support should be contingent on rectifying these wrongs. (WHO 2017a, p. 32) This point would suggest that when the compliance condition is significantly violated, the obligation to support surveillance might cease until the extent of non-compliance recedes. Finally, Article 44 of the WHO International Health Regulations urges, but does not require, wealthier countries to support required surveillance capacities in impoverished areas (WHO 2016). These capacities are aimed to detect potential health emergencies of international concern, not information that might be of use in addressing local health inequities.

Another possibility in situations of widespread non-compliance is that the obligations of others to address inequities are enhanced. On this view, if many are looking the other way, perhaps even corruptly, about the existence of deeply problematic inequities, others might have stronger obligations to step in and help. These might be seen as either a change in duties or as charitable considerations that are not obligatory but are supererogatory. The analysis also might be different depending on whether the actor is a professional, an international governmental organization or a state, or a non-governmental organization.

An enlightening example to explore the obligations of professionals under circumstances of non-compliance might be Dr. Mona Hanna-Attisha, the pediatrician in Flint, Michigan, who exposed the town's water crisis. Hanna-Attisha spent countless hours and endured significant frustration in calling attention to the elevated lead levels she was seeing in her patients. (Hanna-Attisha 2018) For her efforts, she has been honored as a hero, receiving multiple awards (AWIT 2019). Heroism is supererogatory, not what ordinary people are expected to do.

On the other hand, Hanna-Attisha's own view is that she just did what she was obligated to do as a physician caring for her patients: "As physicians we have taken an oath to stand up as the healers and the protectors. We were fighting for the future that lives and grows inside our children and this was not a fight we could lose. Not on my watch." (Hanna-Attisha 2016) On this view, as a physician she had special obligations to go further than others in countering the effects of non-compliance.

This issue of the special obligations of health care providers when others are not doing their part has been discussed extensively in the context of individual risks of exposure to potentially fatal diseases, such as COVID-19, Ebola or SARS (e.g. Ruderman et al. 2006). Positions on this vary, from the view that undertaking a professional role incurs obligations to treat without exception, to the view that people with special obligations to others such as children have competing duties, to the view that undertaking significant personal risks should be up to the individual professional. An additional consideration is fairness in undertaking such special obligations, so that professionals have an obligation to take up their fair share of the slack, but not more. Hanna-Attisha's risks were not to her life, but to her reputation, her health due to stress, and her family life. She saw herself as doing what a good pediatrician would do, but on more limited views of professional obligations she was going far beyond the call of duty.

Many of the possible surveillance actors in circumstances of non-compliance are international organizations, states, or a variety of non-state actors. Here, one question is whether governmental organizations should augment rather than reducing surveillance aid in circumstances of non-compliance. As indicated above, the WHO Guidelines suggest the reverse—that efforts to support surveillance may be suspended at least when non-compliance is sufficiently serious that it violates human rights. The Guidelines also invoke reciprocity (e.g. p. 43), so that if others are not doing their part, obligations of aid may be suspended rather than enhanced. The role of governmental organizations in furthering international cooperation might suggest that reciprocity is an important consideration for them as well.

Non-state actors with extensive resources such as the Gates Foundation have taken up a good deal of the space in international aid efforts. Whether any conclusions should be drawn from this about their obligations is disputed. Some contend that such organizations are voluntary and charitable and so should be permitted to select their own priorities. Others argue that they are subject to obligations such as the obligations of equity. (E.g. Beliuz 2015; Editorial 2009; Wadman 2007).

On such a contrary view, equity arguably requires non-state actors to do more rather than less for those subject to the greatest inequities when others are non-compliant. This view rests squarely on equity to those who are most severely disadvantaged by the non-compliance, particularly children and others subject to corrosive disadvantage. This equity argument might be thought to apply with particular force to certain non-state actors because of the purposes they have undertaken and the roles they have come to play. On this view, non-state actors might be subject to other problems of partial compliance such as feasibility but would continue to have special obligations of equity because of the role they play. We consider these issues about the role of non-state actors further when we discuss the use of new sources of data and new actors in the next chapter.

4.7   Surveillance Under Feasibility Challenges

Feasibility is another important partial compliance consideration: the moral landscape of what ought to be done might shift depending on what may be achieved under existing circumstances. Feasibility might apply either to conducting the surveillance itself or to bringing about any hoped-for change as a result. The role of corruption in blocking change, discussed in the preceding section, is a feasibility consideration, albeit one that arises because of widespread wrongful behavior. In the preceding section, we considered the import of the wrongfulness of the behavior; here, we attend to the role of feasibility more directly, considering primarily the feasibility of surveillance itself. An important caution here is that judgments about feasibility are complex and may involve assumptions about what people are willing to do and what resources they are willing to commit. Judgments of this kind are not judgments about capability *per se*, but rest on further normative considerations (Southwood 2018).

The humanitarian medical organization Médecins Sans Frontières (MSF) illustrates how threats to health care workers can adversely affect surveillance. In the spring of 2019, in the midst of an uncontrolled Ebola epidemic in the Democratic Republic of Congo, MSF was forced to suspend activities at two Ebola treatment centers that had been attacked (MSF 2019a). MSF continued to support treatment facilities elsewhere in the area and to attempt to shore up surveillance to identify new cases—while reminding the world that it was still searching for three MSF health care workers who had been abducted in the region six years earlier. But it was unwilling to subject its workers to further serious risk when they, their facilities, and the patients they served might be destroyed. This example comes close to incapability; dead workers cannot gather data.

In other circumstances, however, strategic efforts may address apparent infeasibility. MSF provides examples of such strategies too. In the Rohingya refugee camps in Bangladesh, where 650,000 people now live, MSF's first effort on entering the camp was to assess the water supply. They found that drinking water was being drawn from pumps located next to latrines and sewage was flowing directly into the aquifer that provided water. MSF's initial step to address this potential health crisis in the camp was to distribute water filters to the most vulnerable, young children and people with compromised immune systems, as a temporary step until safer water sources could be established. MSF then drilled bore holes so that water could be drawn from a source other than the surface aquifer being polluted from the latrines. These strategies of disease prevention were emergency steps, even as other efforts were being pursued to vaccinate camp residents against contagious diseases (MSF 2019b). The video telling this story ends with a note that feasibility depends on context: the cautious observation that these efforts were possible in the camps in Bangladesh but might not be possible elsewhere.

Addressing the effects of global climate change is sometimes met by a general feasibility objection: we have already done so much damage that remediation would be fruitless. Critics of such expressions of despair rightly point out that while some strategies may be foreclosed—continued habitability of low-lying islands such as Nuatambu and others of the Solomon Islands, or of coastal areas such as those in Bangladesh—other strategies remain possible. These strategies, too, may involve assessment of water. For example, in some areas of Bangladesh farmers have been able to adapt to water level changes by diversifying economic activity from farming to fishing rather than by migrating. Where increased salinity of the water has also occurred, however, relocation is more likely. (Chen and Mueller 2018) Feasibility may suggest changes in what is surveilled or how surveillance is conducted, but not in whether to surveil in the first place.

Feasibility concerns have also been raised about whether to communicate the results of surveillance to affected individuals. The WHO Guidelines hold that "results of surveillance must be effectively communicated to relevant target audiences" (WHO 2017a, p. 42). The Guidelines also raise questions about whether surveillance results should always be conveyed, particularly when there is nothing that the person can do with the information and when the information is likely to be stigmatizing or otherwise damaging. They urge a balance of considerations: feasibility, the possibility of taking action, and the potential benefit to the individual. In the judgment of the Guidelines, the balance should be struck in favor of communication, except in rare cases when significant harm might be caused. Anti-paternalism and respect for persons support this conclusion: people should be able to know what surveillance reveals about what has happened to them, even when there is little or nothing to be done to improve their conditions. At a minimum, such information may empower communities to try to take steps to prevent further harms than the ones they have already experienced. Failure to provide this information would add another inequity to communities that are already experiencing significant other inequities.

4.8   Water Surveillance and Ideal Surveillance

Another partial compliance concern is the relevance of ideals to actions in non-ideal circumstances. Should we decide what to do about surveillance in the here-and-now by developing a model of what good surveillance would look like, and engage in planning in the here-and-now to try to move more closely to that goal? A "yes" answer to this question faces several challenges: developing a model of ideal surveillance in the first place, ascertaining what at any given point in time would count as progress towards it, and considering other consequences of such supposedly progressive choices. We begin with doubts about the project of developing a model of ideal surveillance.

Proposals for ideal surveillance systems that have been advanced press questions about what makes surveillance ideal. Bentham's Panopticon—all seeing and all knowing—would obtain all information about everyone, instantly and forever. Such perfect surveillance has met with immediate objections: awareness of the existence of constant observation would threaten individual privacy, freedom of thought, autonomy, identity, and sense of self. Ideal surveillance, it would seem, is not the same as omnipresent information-gathering; rather, if it is to be ideal, it must take its place among other goods in human life.

Environmental surveillance does not yield immediately identifying information about individuals, however. Measuring the water quality in an area is not the same as measuring the effect of drinking that water on the health of an individual. So perhaps ideal environmental surveillance could be complete information in a way that would not be threatening to individuals' other values. One answer to this objection is that it may not be easy to separate surveillance of the environment from surveillance of individuals within the environment. As we described earlier in this chapter, environmental surveillance may lead to more intrusive forms of surveillance. Another answer is that complete information is not the same as usable information, and choices will still need to be made in terms of relevance and organization about whatever information is to be collected. We thus have doubts about the enterprise of constructing an account of an ideal surveillance system, even for environmental matters such as water quality.

Setting these doubts aside, however, there are also problems about whether progress towards more ideal surveillance should be the primary aim when circumstances are less than ideal. What if efforts to gather more complete information needed to improve health would threaten some groups? Can more information be bad and, if so, what should be done about getting it? The WHO Guidelines for ethical surveillance recognize these concerns, using data that might reveal stigmatized behavior as an example (WHO 2017a, p. 27).

The Guidelines' response to this difficulty is procedural: communities should be engaged in creating and implementing oversight mechanisms to identify risks and benefits of surveillance for themselves. According to the Guidelines, these oversight mechanisms should be chosen in a manner that is transparent and accountable to the communities concerned. The oversight mechanisms should try to ensure that information gathered for surveillance is only used for public health purposes and not for other purposes that could threaten individuals in the community by revealing confidential information about them (WHO 2017a, p. 29). Oversight mechanisms should also continuously monitor for risks of harm (WHO 2017a, p. 34).

Nonetheless, the WHO Guidelines also recognize that harms may be unavoidable. When serious health emergencies are possible, surveillance may be necessary whatever the community's views about potential harms to them. In such circumstances, the Guidelines urge efforts to identify risks in advance and mitigate them where possible. Mitigation efforts should include protections against harm to those who surveil and those who report illness, including health care workers and humanitarian organizations. When harms prove inevitable, mitigation efforts should include compensation. (WHO 2017a, p. 34) Examples of compensation are sick pay for those deprived of work or payments for culled poultry stock. Moreover, according to the WHO Guidelines, when risks of harm cannot be eliminated, they should be proportional to the benefits to be gained from surveillance. (WHO 2017a, p. 35). Another ethical consideration is whether the benefits of surveillance accrue to those experiencing the harm.

Experience with mitigation efforts is not always encouraging, however. After cases of avian influenza were identified in West Bengal, flocks were culled but the compensation received did not come close to making up for the economic losses, let alone the losses to communities that resulted from the culling (Chakraborti 2009). In the United States, homeowners in areas identified with environmental toxicities have found themselves with steep losses in housing values (Currie et al. 2015). Housing values in Flint had been on the upswing before the water crisis but fell sharply in its aftermath (Goldstein 2016).

Overall, the approach of the WHO Guidelines to risks of improved surveillance can be characterized as oversight, transparency, and community involvement. But there are difficulties in relying on these procedural mechanisms in some partial compliance contexts. The most obvious are barriers to creating oversight mechanisms in the first place. An extensive literature addresses the problem of failed states where even rudimentary mechanisms of civil society may need to be built from the ground up. MSF (2019c) has an illuminating term for what might be possible in circumstances in which surveillance is not: "Témoignage – translated as bearing witness – is the act of raising awareness, either in private or in public, about what we see happening in front of us."

Another problem is trying to understand what community thoughts, reactions, or preferences might mean and how we might assess the community's voice. We consider in Chapter 7 how community may be defined and engaged. Here, we point out the role of equity concerns in such circumstances. Community choices may vary in the extent that they attend to the needs of the most vulnerable, especially those who cannot speak for themselves. Children do not participate directly in the political process and must rely on others for adequate representation of their interests. As we noted at the beginning of this chapter, environmental issues are complicated by the possibility of free riders, individuals who are able to experience the benefits of public goods without themselves bearing any of the costs of their production. Procedures for gathering community input may find it difficult to counter the incentives of these free riders. Communities may also seek to avoid costs to themselves from environmental hazards while being willing to allow the risks to occur in the backyards of other communities.

Most importantly, efforts to address equity should focus on those who are most vulnerable, those who cannot speak for themselves and who may be subject to what Wolff and De-Shalit (2007) called "corrosive" disadvantage. Corrosive disadvantages are disadvantages that interact, creating even greater disadvantages than either would on its own. For example, exposure to poor environmental conditions such as lead in the water might interact with inadequate educational systems to result in downward spiraling lives for children growing up in these circumstances. Corrosive disadvantages in turn may make it more difficult for people who are subject to them to participate adequately in community decision making.

4.9   Summary

If a goal of public health is health equity, equity in surveillance is imperative. Surveillance equity requires attention to what is ignored as well as to what is more immediately salient. Environmental surveillance is complicated by incentives to free ride and to allow hazards to occur elsewhere so long as there are no apparent risks of spread. In an unjust world, efforts to address equity must attend most urgently to corrosive disadvantages and their impact on those who are least able to protect themselves.

References

Adamson, Adewole S., and Avery Smith. 2018. Machine Learning and Health Care Disparities in Dermatology. *JAMA Dermatology* 154 (11): 1247–1248.

Allaire, Maura, Haowei Wu, and Upmanu Lall. 2018. National Trends in Drinking Water Quality Violations. *Proceedings of the National Academy of Science* 115 (9): 2978–2083.

Atkin, Emily. 2018. Did Flint's Water Crisis Damage Kids' Brains? *The New Republic* (Feb. 14) [online]. https://newrepublic.com/article/147066/flints-water-crisis-damage-kids-brains. Accessed 20 July 2020.

AWIT. 2019. *Dr. Mona Hanna-Attisha.* https://www.americanswhotellthetruth.org/portraits/dr-mona-hanna-attisha. Accessed 20 July 2020.

Balazs, Carolina L., and Isha Ray. 2014. The Drinking Water Disparaties Framework: On the Origins and Persistence of Inequities in Exposure. *American Journal of Public Health* 104 (4): 603–611.

Beliuz, Julia. 2015. The Media Loves the Gates Foundation. These Experts are More Skeptical. *Vox* (June 10) [online] https://www.vox.com/2015/6/10/8760199/gates-foundation-criticism. Accessed 20 July 2020.

Braveman, Paula, Elaine Arkin, Tracy Orleans, Dwayne Proctor, and Alonzo Plough. 2017. *What is Health Equity? And What Difference Does a Definition Make?* Princeton: Robert Wood Johnson Foundation.

Braveman, Paula, Julia Acker, Elaine Arkin, Dwayne Proctor, Amy Gillman, Kerry Anne McGeary, and Giridhar Maliya. 2018. *Wealth Matters for Health Equity.* Princeton: Robert Wood Johnson Foundation.

Centers for Disease Control and Prevention (CDC). 2017. *Lead: What Do Parents Need to Know to Protect Their Children.* https://www.cdc.gov/nceh/lead/acclpp/blood_lead_levels.htm . Accessed 20 July 2020.

Chakraborti, Chhanda. 2009. Pandemic Management and Developing World Bioethics: Bird Flu in West Bengal. *Developing World Bioethics* 9 (3): 161–166.

Chen, Joyce, and Valerie Mueller. 2018. Coastal Climate Change, Soil Salinity and Human Migration in Bangladesh. *Nature Climate Change* 8: 981–985.

Currie, Janet, Lucas Davis, Michael Greenstone, and Reed Walker. 2015. Environmental Health Risks and Housing Values: Evidence from 1,600 Toxic Plant Openings and Closings. *American Economic Review* 105 (2): 678–709.

Drum, Kevin. 2016. Raw Data: Lead Poisoning of Kids in Flint. *Mother Jones* (Jan. 25) [online]. https://www.motherjones.com/kevin-drum/2016/01/raw-data-lead-poisoning-kids-flint/. Accessed 20 July 2020.

Editorial. 2009. What has the Gates Foundation Done for Global Health? *The Lancet* 373 (9675): 1577.

Egan, Paul. 2017. These Are the 15 People Criminally Charged in the Flint Water Crisis. *Detroit Free Press* (June 14) [online]. https://www.freep.com/story/news/local/michigan/flint-water-crisis/2017/06/14/flint-water-crisis-charges/397425001/. Accessed 20 July 2020.

Erb, Robin. 2015. Flint Doctor Makes State See Light About Lead in Water. *Detroit Free Press* (Oct. 10) [online]. https://www.freep.com/story/news/local/michigan/2015/10/10/hanna-attisha-profile/73600120/. Accessed 20 July 2020.

European Environmental Agency (EEA). 2018. *European Waters: Assessment of Status and Pressures 2018*. EEA Report No. 7/2018. https://cdn.downtoearth.org.in/uploads/european-waters-assessment2018.pdf. Accessed 20 July 2020.

Goldstein, Daniel. 2016. Lead Poisoning Crisis Sends Flint Real-Estate Market Tumbling. *Market Watch* [online] (Feb. 17). https://www.marketwatch.com/story/lead-poisoning-crisis-sends-flint-real-estate-market-tumbling-2016-02-17. Accessed 20 July 2020.

Grossman, Daniel S., and David J.G. Slusky. 2017. *The Effect of an Increase in Lead in the Water System on Fertility and Birth Outcomes: The Case of Flint, Michigan* (Aug. 17). West Virginia University Department of Economics Working Paper No. 17–25. http://busecon.wvu.edu/phd_economics/pdf/17-25.pdf. Accessed 20 July 2020.

Hanna-Attisha. 2016. *Flint's Fight for America's Children, TEDMED Talk*. https://www.tedmed.com/talks/show?id=627338. Accessed 20 July 2020.

Hanna-Attisha, Mona. 2018. *What the Eyes Don't See: A Story of Crisis, Resistance, and Hope in an American City*. New York: One World Random House.

Hanna-Attisha, Mona, Jenny LaChance, Richard Casey Sadle, and Allison Champney Schnepp. 2016. Elevated Blood Lead Levels in Children Associated With the Flint Drinking Water Crisis: A Spatial Analysis of Risk and Public Health Response. *American Journal of Public Health* 106 (2): 283–290.

Highsmith, Andrew R. 2015. *Demolition Means Progress: Flint, Michigan, and the Fate of the American Metropolis*. Chicago: University of Chicago Press.

Institute of Health Equity. 2010. *Fair Society, Healthy Lives: The Marmot Review*. http://www.instituteofhealthequity.org/resources-reports/fair-society-healthy-lives-the-marmot-review/fair-society-healthy-lives-exec-summary-pdf.pdf. Accessed 20 July 2020.

Integrative Human Microbiome Project Research Network Consortium (iHMP Consortium). 2019. The Integrative Human Microbiome Project. *Nature* 569: 641–648.

Mack, Elizabeth A., and Sarah Wrase. 2017. A Burgeoning Crisis? A Nationwide Assessment of the Geography of Water Affordability in the United States. *PLoS One* 12 (1): e0169488. https://journals.plos.org/plosone/article?id=10.1371/journal.pone.0169488. Accessed 20 July 2020.

McDonald, Yolanda J., and Nicole E. Jones. 2018. Drinking Water Violatioins and Environmental Justice in the United States, 2011–2015. *American Journal of Public Health* 108 (10): 1401–1407.

Médecins Sans Frontières (MSF). 2019a. *Ebola Response Failing to Gain the Upper Hand on the Epidemic*. https://www.msf.org/ebola-response-failing-gain-upper-hand-epidemic-democratic-republic-congo. Accessed 20 July 2020.

———. 2019b. How We Work: *5 Minutes to Explain* (video). https://www.msf.org/how-we-work. Accessed 20 July 2020.

———. 2019c. *How We Work*. https://www.msf.org/how-we-work. Accessed 20 July 2020.

Michigan Civil Rights Commission. 2017. *The Flint Water Crisis: Systemic Racism Through the Lens of Flint* (Feb. 17). http://www.michigan.gov/documents/mdcr/VFlintCrisisRep-F-Edited3-13-17_554317_7.pdf. Accessed 20 July 2020.

Mota, Caitlin. 2016. More Jersey City Parks Closed After Discovery of Lead Chips, Official Says. *The Jersey Journal* (March 20) [online]. http://www.nj.com/hudson/index.ssf/2016/03/more_jersey_city_parks_closed_after_discovery_of_l.html. Accessed 20 July 2020.

Rawls, John. 1971. *A Theory of Justice*. Cambridge, MA: Harvard University Press.

Rothstein, Richard. 2017. *The Color of Law: A Forgotten History of How Our Government Segregated America*. New York, NY: Liveright Publishing.

Ruderman, Carly, C. Shawn Tracy, Cécile M. Bensimon, Mark Bernstein, Laura Hawryluck, Randi Zlotnik Shaul, and Ross E.G. Upshur. 2006. On Pandemics and the Duty to Care: Whose Duty? Who Cares? *BMC Medical Ethics* 7: 5. https://doi.org/10.1186/1472-6939-7-5. Accessed 20 July 2020.

Sen, Amartya. 2011. *The Idea of Justice*. Cambridge, MA: Harvard University Press.

Sokolow, Susanne H., Isabel J. Jones, Merlijn Jocque, Diana La, Olivia Cords, Anika Knight, Andrea Lund, Chelsea L. Wood, Kevin D. Lafferty, Christopher M. Hoover, Phillip A. Collender, Justin V. Remais, David Lopez-Carr, Jonathan Fisk, Armand M. Kuris, and Giulio A. De Leo. 2017. Nearly 400 Million People are At Higher Risk of Schistosomiasis Because Dams Block the Migration of Snail-Eating River Prawns. *Philosophical Transactions of the Royal Society of London B: Biological Sciences* 372 (1722): 20160127.

Southwood, Nicholas. 2018. The Feasibility Issue. *Philosophy Compass* 13 (8): e12509.

Sultana, Farhana. 2018. Water Justice: Why it Matters and How to Achieve It. *Water International* 43 (4): 483–493.

Swarns, Rachel. 2015. Biased Lending Evolves, and Blacks Face Trouble Getting Mortgages. *The New York Times*. (Oct. 30) [online]. https://www.nytimes.com/2015/10/31/nyregion/hudson-city-bank-settlement.html . Accessed 20 July 2020.

Tiemann, Mary. 2017. Safe Drinking Water Act (SDWA): A Summary of the Act and Its Major Requirements. *Congressional Research Service* (March 1). https://fas.org/sgp/crs/misc/RL31243.pdf. Accessed 20 July 2020.

Tolins, Molly, Mathuros Ruchiriwat, and Philip Landrigan. 2014. The Developmental Neurotoxicity of Arsenic: Cognitive and Behavioral Consequences of Early Life Exposure. *Annals of Global Health* 80 (4): 303–314.

UN Water. 2019. *Water and Climate Change*. http://www.unwater.org/water-facts/climate-change/. Accessed 20 July 2020.

United Nations (UN). 2018a. *Clean Water and Sanitation: Why it Matters*. https://www.un.org/sustainabledevelopment/wp-content/uploads/2018/09/Goal-6.pdf. Accessed 20 July 2020.

———. 2018b. *The Sustainable Development Goals Report*. https://unstats.un.org/sdgs/files/report/2018/TheSustainableDevelopmentGoalsReport2018-EN.pdf. Accessed 20 July 2020.

United Nations Environmental Programme. 2017. *UN Environment Assembly Moves to Curb Pollution from Extractive Industries* (Dec. 14). https://www.unenvironment.org/news-and-stories/story/un-environment-assembly-moves-curb-pollution-extractive-industries. Accessed 20 July 2020.

United States Census Bureau. 2019. Quick Facts: Flint city, Michigan. https://www.census.gov/quickfacts/flintcitymichigan. Accessed 19 February 2021.

United Nations General Assembly. 2010. *Resolution 64/292. The Human Right to Water and Sanitation* (Aug. 3). https://undocs.org/A/RES/64/292. Accessed 20 July 2020.

Valentini, Laura. 2012. Ideal vs. Non-Ideal Theory: A Conceptual Map. *Philosophy Compass* 7 (9): 654–664.

VanDerslice, James. 2011. Drinking Water Infrastructure and Environmental Disparities: Evidence and Methodological Considerations. *American Journal of Public Health* 101 (Suppl. 1): S109–S114.

Wadman, Meredith. 2007. State of the Donation. *Nature* 447: 248–250.

Water Integrity Network (WIN). 2016. *Water Integrity Global Outlook* 2016. https://www.water-integritynetwork.net/. Accessed 20 July 2020.

Wines, Michael, and John Schwartz. 2016. Unsafe Lead Levels in Tap Water Not Limited to Flint. *The New York Times* (Feb. 8) [online]. https://www.nytimes.com/2016/02/09/us/regulatory-gaps-leave-unsafe-lead-levels-in-water-nationwide.html. Accessed 20 July 2020.

Wines, Michael, Patrick McGeehan, and John Schwartz. 2016. Schools Nationwide Still Grapple With Lead in Water. *The New York Times* (March 26) [online]. https://www.nytimes.com/2016/03/27/us/schools-nationwide-still-grapple-with-lead-in-water.html?action=click&contentCollection=U.S.&module=RelatedCoverage&region=Marginalia&pgtype=article. Accessed 20 July 2020.

Wolff, Jonathan. 2017. Forms of Differential Social Inclusion. *Social Philosophy and Policy* 34 (1): 164–185.

Wolff, Jonathan, and Avner De-Shalit. 2007. *Disadvantage*. Oxford: Oxford University Press.

World Commission on Dams. 2000. *Dams and Development: A New Framework for Decision-Making*. https://www.internationalrivers.org/sites/default/files/attached-files/world_commission_on_dams_final_report.pdf. Accessed 20 July 2020.

World Health Organization (WHO). 2016. *International Health Regulations (2005)*, 3d ed. https://www.who.int/ihr/publications/9789241580496/en/. Accessed 20 July 2020.

———. 2017a. *WHO Guidelines on Ethical Issues in Public Health Surveillance*. Geneva: World Health Organization. Licence: CC BY-NC-SA 3.0 IGO. https://apps.who.int/iris/bitstream/handle/10665/255721/9789241512657-eng.pdf;jsessionid=064C310DEE572FB61E564C7A1FA172D7?sequence=1. Accessed 20 July 2020.

———. 2017b. *Progress on Drinking Water, Sanitation and Hygiene*. http://apps.who.int/iris/bitstream/handle/10665/258617/9789241512893-eng.pdf;jsessionid=A6237201F3370F975ADEC65C46CFD347?sequence=1. Accessed 20 July 2020.

———. 2018. *Diarrhoeal Disease*. http://www.who.int/en/news-room/fact-sheets/detail/diarrhoeal-disease. Accessed 20 July 2020.

———. 2019. *IHR Core Capacities Implementation Status: Surveillance*. https://www.who.int/gho/ihr/monitoring/surveillance/en/. Accessed 20 July 2020.

World Health Organization Commission on the Social Determinants of Health. 2008. *Closing the Gap in a Generation: Health Equity Through Action on the Social Determinants of Health*. https://apps.who.int/iris/bitstream/handle/10665/43943/9789241563703_eng.pdf;jsessionid=DF20E8A35BBA43D30CEEB69BDF16FDCE?sequence=1. Accessed 20 July 2020.

Chapter 5
Enhancing Surveillance: New Data, New Technologies, and New Actors

5.1 Introduction

In many areas of the globe, the capacities of public health to gather information about health are rudimentary at best. Health care, transportation, power, and communication infrastructures may be very limited, if not entirely absent. In some of these areas, new technologies are filling gaps. To take perhaps the most impressive example, cell phone and even smart phone access is increasing rapidly using local solar power and serving areas without hard infrastructure. Data collected through these mechanisms can be critical sources of information about users, their health, their locations, and their activities.

In wealthier nations, too, surveillance now uses sources that go far beyond surveys conducted by public health agencies and data collected through the reporting requirements of state and national laws. Electronic medical records, patient registries, research data bases, direct to consumer testing, smart phone apps, internet searches, social media sites, webcams, and smart devices making up the "internet of things" all contribute highly useful information about health. Much of this is used by public health, and much more is used by private entities to survey the health of the public.

These data sources are far richer in the information they contain and may also be more representative of the population than are sources available to traditional public health. Combining data from these sources may also reveal unexpected patterns and correlations that can be especially helpful for syndromic surveillance, a recently developed form of surveillance that uses artificial intelligence to identify patterns that may signal disease clusters or unusual events. Devices such as smart phones may give location information in addition to content that can enable detection of hot spots for exposures. The information may be available in real time, thus enabling quick responses to emergencies. The use of existing data sources also can avoid the redundancy and expense of efforts by public health to collect its own data.

© Springer Nature Switzerland AG 2021
J. G. Francis, L. P. Francis, *Sustaining Surveillance: The Importance of Information for Public Health*, Public Health Ethics Analysis 6,
https://doi.org/10.1007/978-3-030-63928-0_5

But there are disadvantages and risks to these innovations (e.g. Kostkova 2018). Data initially collected for purposes other than public health may not meet the standards of public health. Information may be collected in different formats that are hard to link together and that risk reduplication or misidentification of individuals. It may contain unexpected inaccuracies or selection biases. The data collected for another purpose may be partially or imperfectly relevant for public health. People may be surprised to find that data have been collected and used, even when the use is for public health purposes of which they would otherwise have approved. Data that have public health uses may also be put to other purposes that people regard as less acceptable, such as marketing, development of expensive for-profit drugs, or criminal investigations. People may be outed, doxed, shamed, harassed, bullied, arrested, deported, or subjected to other risks when data are drawn from other sources to be used by public health. The power of these new technologies allows far more to be inferred or known about individuals, their movements to be traced, and quite complete portraits of them to be created—all, perhaps, without their knowledge or participation in any way. Anger, protests, and even efforts to have the data destroyed may be the result.

Technology is evolving so rapidly that it would be foolhardy to assume current uses will continue to predominate or that likely new data sources are predictable. Consequently, the discussion in this chapter uses important illustrations drawn from what is happening with data today rather than claiming to be comprehensive. In what follows, we describe these novel data sources, many of which are in private hands or use techniques of artificial intelligence for analysis:

– Interoperable electronic health records
– Retained blood spots from newborn screening
– Biobanks and other genetic databases
– Patient registries
– Information gained in research
– Direct to consumer testing, including genetic testing
– Smartphones and smartphone apps
– Wearables (e.g. fitbit, AppleWatch) and biosensors
– Robots and smart devices

Data from some or all of these sources are increasingly being linked, creating vast reservoirs of data.

Novel actors are also entering the surveillance space. These include non-state partners with WHO and health care providers. They also include internet search engines such as Google and social media sites such as Facebook. While these may also be both sources and users of data for many purposes, the information they collect and some of the uses they make of that information may be beneficial for public health. These business entities also have commercial interests that may be at odds with the goals of public health.

Cutting across some of these new data and new actors are features that are particularly important to the core considerations we outlined at the beginning of this volume and that we think are critical to sustain trust in surveillance. Many involve "big data": vast data reservoirs that are huge in volume, velocity and variety. These data are rich resources for the development of analytic and predictive algorithms for the health of populations and subpopulations, communities and groups, and families and individuals. Although the data often are initially de-identified, they may be combined in ways that significantly increase the risks that individuals may be re-identified or that inferences may be drawn about individuals. Inclusion in these data reservoirs may be without notice or consent at all, with opt-out or opt-in methods of agreement, or by "broad" consent that may stretch for years or even lifetimes.

5.2 "Big" Health Data and AI

Technological abilities to produce, collect, store, and analyze data are growing exponentially. Discussions of the ethics of big data are also growing apace (e.g. Mittelstadt and Floridi 2016). In this section and the next two, we focus on three of these overarching ethical issues with novel data sources of particular relevance to public health surveillance: artificial intelligence analytics and bias, identification of or inferences about individuals, and notice and consent.

What makes data "big" is itself subject to debate. Mittelstadt and Floridi, in a meta-analysis of publications on the ethics of big data, note that definitions include the numbers of individuals in the data base, the sheer amount of data, the computing technology needed to analyze the data, and the types of analyses that can be conducted. "Big" data does not mean data that are complete, accurate, or without bias, however. To take just one example, critics note that big data assembled from electronic medical records may include errors in data entry and selection bias in what is entered (Hoffman and Podgurski 2013).

Predictive analytics may be applied to these data sets. In what is termed a "learning health care system," for example, algorithms may be developed that predict patient outcomes from the data. As more and more data are accumulated, the algorithms may be refined for greater and greater predictive accuracy, enabling the health care system to refine care delivery. One of the issues with the use of these predictive algorithms is that the patterns that they identify in the data may not be predictable in advance. Thus with techniques such as syndromic surveillance it may only be possible to tell people after the fact what was discovered. While some goals can be outlined—improving health or detecting drug side effects, for example—any precision about what might be linked to these goals may be unknown. Thus models of informed consent that require disclosures of possible risks may not be possible (Francis et al. 2009).

These predictive algorithms are used not only in health care but also in areas that may have differential impacts on the health of members of a population: for example, predicting whether someone is a good credit risk (and thus can qualify for a loan or a mortgage), a reliable tenant (and thus can be rented an apartment in a competitive market), or a likely re-offender (and thus should not be placed on parole). Facebook has used algorithms to predict suicide risk and google to predict influenza outbreaks—with, as we describe below, success that has been uneven.

Bias can be introduced at many levels in this process. The data itself are an initial problem. Critics of the use of predictive algorithms in policing point out that if police are deployed primarily in minority neighborhoods, and drug arrests are higher in those neighborhoods, this does not mean that drug use is actually higher in those neighborhoods. It only means that data have been gathered from those neighborhoods (e.g. Ferguson 2017). Similarly, if biobanks include primarily Caucasians of Northern European descent, algorithms will not "learn" from data about persons with origins in which other genotypes are more common.

Algorithms used in predictive analytics are often not transparent; in the words of Frank Pasquale (2016), their use has created a "black box" society. Commercial entities developing these techniques argue that they should be protected as trade secrets (Meyers 2019). The ability to scrutinize bias in their development and application is limited by these protections.

Even if these algorithms were public, there is controversy about what it would mean for an algorithm to be "fair." Friedler and colleagues (2016) argue that "fairness" has been understood in fundamentally different ways. Algorithms link decisions (for example, what treatment to recommend for a particular patient) to constructs (for example, a score about likelihood of survival) based on observed proxies for the construct (for example, heart ejection fraction or kidney function). Bias can come in at any of these points, including the observations collected, what are determined to be proxies for the construct, and whether the construct is appropriately linked to the decision space. With recidivism, to take an example from criminal law, prior arrest record might be an observation used as a proxy for the person's "propensity" to commit a crime, which is then used to predict the likelihood that someone will reoffend and therefore should not be let out on parole. There may be structural bias in whether the construct space is appropriately represented by the observed space as a proxy. This may be skewed by groups such as race, sex, or age as well (Zou and Schiebinger 2018).

A further justice problem presented by AI is the "divide" between data subjects and those with the resources to analyze the data (Benkler 2019). Data may be scraped from all kinds of sources used by ordinary people, from grocery store discounts to medical records. However, big data collection, maintenance, and analysis take massive resources. If data are repurposed and used for gains of others—as they might be for decisions such as whether to grant credit, whether to rent housing, how to treat patients who can access care, or where to locate a primary care clinic—distrust might be the predictable result (Francis and Francis 2014).

5.3   The Debate About Re-identification

A standard distinction in data protection law lies between data that can be used to identify individuals and data that have been de-identified. The GDPR (2020), the U.S. rules governing research with human subjects, and the U.S. rules governing the privacy and security of health information possessed by covered health care entities and health insurers all apply only to identifiable information. Once information has been appropriately de-identified—and it must be emphasized that these standards are stringent—the only further regulatory issue is that it not be re-identified. Data use agreements may prohibit re-identification efforts but ironically by so doing may also rule out external assessments of whether the data are at risk of re-identification (Sweeney et al. 2018). Privacy policies and terms and conditions from websites and social media sites typically draw this line as well. But the line has become increasingly controversial. Possibilities of re-identification pose one set of problems. Possibilities of harms to groups and harmful inferences about individuals pose another.

As data sets are increasingly combined, information becomes richer and richer. The more data are enriched, the greater the likelihood that properties will combine so that individuals ultimately stand out. The Harvard computer scientist Latanya Sweeney has conducted a number of well-known re-identification "attacks" (e.g. Yoo et al. 2018; Sweeney et al. 2018; *Southern Illinoisan* 2006). While some of these attacks required sufficient sophistication for Illinois courts to decide that the release of information collected by the state about cancer patients did not put individual privacy at risk, others have become increasingly easy. One study also indicates that reidentification of law students may be more likely if the students are Black or Hispanic, thus introducing the possibility of disparate impact discrimination in reidentification risk (Sweeney et al. 2018).

Identity-masking techniques have been developed to help to disguise individual identities. Some forms of identity masking involve taking out sufficient amounts of information so that individuals cannot be mapped to cells containing very small numbers. An example is the U.S. HIPAA privacy rule which presumes that data are deidentified if a list of eighteen kinds of information are removed. Other forms introduce noise to the extent that differential privacy is achieved in the sense that it is not possible to tell whether any given individual was included in the data set. The choice of anonymization techniques has costs, however. Removing information or introducing noise may reduce the utility of the data set, especially if this is done to the extent needed to reduce reidentification risks in light of current reidentification technologies. Xu and Zhang (2020) argue in addition that the choice of deidentification techniques may affect the extent to which disparities are observed from the data that could support claims of disparate impact discrimination. If noise insertion is the anonymization technique, and disparities are understood as the separation between groups, it is less likely that disparities will be identified because noise insertion may tend to increase the standard deviation of outcomes within population subgroups.

Even when de-identification is achieved successfully, further ethical concerns remain. When individuals can be mapped successfully to groups that are sufficiently small to allow inferences to be drawn about them with high probability, individuals may suffer stigmatization, shaming, or other deleterious consequences. Group-level harms may also occur, as when the conclusions about tribal origins were drawn about the Havasupai from the research described later in this chapter. Moreover, the economic incentives may be much greater to develop understanding about the characteristics of groups rather than individuals: "high-payoff" targets for commercial actors are not single individuals but individuals as members of groups (Floridi 2014).

5.4   The Absence of Real-Time Notice or Consent

Further ethical issues are posed by whether individuals have notice of or can in any way be said to consent to data collection or use. If de-identified data are considered no longer to be about the individual, as is standard in legal regimes governing data protection, the conclusion is drawn that individuals need not be given notice. Standard boilerplate in the terms and conditions for website users, for example, is that information that does not reveal identities may be used or transferred. The Google Privacy Policy is an illustration: "We may share non-personally identifiable information publicly and with our partners—like publishers, advertisers, developers, or rights holders. For example, we share information publicly to show trends about the general use of our services. We also allow specific partners to collect information from your browser or device for advertising and measurement purposes using their own cookies or similar technologies" (Google 2020). The result may be surprise on the part of individuals about how information is being used or transferred, especially if the information is combined with other information that can be used to identify them or draw inferences about them. Individuals may accept this from Google—the service is free and people are used to the advertising they receive as a result—but be distressed if it is used in ways that might affect their health or the health of the public.

Some data users employ "opt in" or "opt out" mechanisms to secure individual agreement to data collection and use. Examples are the ubiquitous "I agree" hot buttons when users seek services over the internet. "Opt in" mechanisms are far less likely to result in user participation because they require positive action, for example signing up to receive content from a newsletter. As discussed below for newborn screening, opt in requirements have greatly reduced participation in letting samples be retained for further public health uses. On the other hand, the U.K. collection of patient data from the National Health has chosen to use an opt out model to increase the extent to which the data are representative of population health (Understanding Patient Data 2018). The concern about this choice is that people may remain largely unaware about collection and use of information drawn from their health care.

"Broad consent" is a controversial strategy for agreement to use of information on an ongoing basis. It is consent, but open-ended. For example, consent to allow medical records to be used in "future health-related research" until revoked is broad consent. This form of consent has been used not only for biobanks but also for registries and many other uses of identifiable information. It has the advantage that individuals do not need to be re-contacted each time a new use of information about them is proposed. Instead, protection of the individuals would rest in any other required approval of the use. For research, the approval might include whether it is health-related and thus within the scope of the consent, whether it appropriately protects individual confidentiality, and whether its benefits outweigh risks. But consent may be far broader. For example, patientslikeme, a site that allows individuals to share their health information, has a trademarked "digitalme ignite" pilot study through which individuals can contribute blood samples, health updates, regular examinations, and other biospecimens. Information is kept indefinitely unless permission is revoked and may be used by researchers or for-profit companies. (patientslikeme 2020).

On the one hand, it seems that individuals should be able to say that they are willing to participate in an ongoing project collecting health information for health-related research or other activities, and to be included in any further research projects as long as their identities are protected. This could include public health research as related to health. Re-contacting individuals is inconvenient and costly and might be regarded as an annoying reminder of health issues.

On the other hand, memories fade. If people are not sent updates about what has been learned from the ongoing data use to which they have consented, they may be surprised and concerned if they hear about it. Commentators have suggested that broad consent is too empty to constitute genuine understanding, that the idea of "health-related" is pliable and may change, and that there should be some kind of expiration, requirement for re-consent, or the opportunity to opt out of continuing at regular intervals. One particularly important moment for re-consent might be adulthood. Some broad consent is given by parents for their children to enter into disease registries as infants and individuals thus included might be very surprised and distressed to discover that information about them has been updated ever since (Francis 2014).

Ploug and Holm (2016) propose "meta consent" as a solution to problems of broad consent. This idea is that people should be asked at the point of data collection to state their preferences about future consent. These choices might be never to be asked, always to be asked, to be asked in some cases or not others, and so on. They argue that this idea far better reflects differences in values: people might want to be asked about some data uses but be fully willing to agree to many other data uses without further contact. They argue that this approach is better able to respond to individual vulnerabilities and sources of mistrust as well. One suggestion that they do not make, but that could also enhance trust, is that when possible any kind of agreement to ongoing data use should be coupled with requirements to provide updates on a regular basis. Such updates might be easily arranged through electronic methods of communication. For example, if information is being collected from electronic health records, as described in the next section, people could be sent updates just as they are sent other communications such as reminders from their health care providers.

5.5 Interoperable Electronic Health Records (EHRs)

Records of individual health care are potentially a very rich source of information for public health. Interoperable EHRs, now widely used in health care, have great advantages for both individual health care and public health. They may be accessible in real time and across long distances, as long as the Internet itself is working and connections can be established. They will likely contain much of the relevant information about the health of individuals who regularly see medical providers. Depending on their design, they are readily searchable and thus have the capacity for public health to follow up and to gather new information as needed. They may be used for syndromic surveillance to identify unusual patterns that might warrant further investigation. They may provide information about rare drug side effects or patient outcomes with experimental medicines; analysis of data in patient medical records is the basis of the sentinel system in the U.S. for monitoring drug safety, for example (Coyle 2017). They may be configured with metadata that record information about the time and source of data collection, thus enabling judgments about reliability or relevance. They likely contain contact information about patients, yet they also may be programmed to de-identify information or to mask particular data fields considered irrelevant or sensitive. They are interactive, allowing alerts or reminders to be delivered to physicians, patients, or others. Use of EHRs may be more efficient for providers to transmit data to public health quickly and accurately than reports given over the telephone or typed into a website. There thus may be considerable pressure to rely on EHR data to supplement or supplant separate collection of data for public health, especially if funding for public health is restricted.

On the other hand, use of EHR data for public health—or at least, reliance on this data to a significant extent in the public health enterprise—presents major challenges even in well-off nations where automated data systems are ubiquitous and information technology is robust. EHR design may not capture data that are important to public health or organize data is a way that is readily accessible for public health purposes. When groups within the population do not have reliable access to health care, EHRs will be unrepresentative of the population as a whole and likely to miss disadvantaged groups. Use of these records for purposes other than patient care may be objectionable to some patients—so much so that they may lose trust in health care and avoid interaction with health care providers as much as they can. Suspicion may intensify with any suggestion that public health information may also be used for law enforcement or security.

Arguably, public health is at least health and many of its activities are devoted to improving health care and population health broadly construed. Nonetheless, access to these records presents serious privacy risks, especially if they can be entered, copied, and downloaded at the flick of a switch without patient agreement or even knowledge. Public health law scholar Wendy Mariner argues that public health use of medical records without individual consent is "mission creep" that threatens individual rights (Mariner 2007). These risks are magnified by the richness of the information within EHRs; even de-identified EHR information may be re-identifiable when combined with other data sources.

5.5.1 EHRs in the United States

In the U. S. today, the vast majority of medical records are electronic. These records are protected by a patchwork of legal standards. These standards include the federal constitution and federal statutory and regulatory protections for covered health information. They also include a variety of state laws.

At the level of the federal constitution, a decision in 1977 held that a state requirement to report controlled substance prescriptions to a state data base did not violate privacy rights (*Whalen* 1977). The reasoning relied on the state's compelling interest in preventing drug abuse and the structuring of the data base to prevent unauthorized disclosures of the information. This is the only Supreme Court decision dealing with data privacy, and it was handed down over forty years ago. The extent to which there is federal constitutional protection for informational privacy remains clouded; one federal appellate court held in 2020 that sexual assault victims could not sue local officials for releasing information that might identify them because the constititutional right to informational privacy was not clearly established (*Dillard* 2020).

The Health Insurance Portability and Accountability Act (HIPAA) is the federal statutory structure that governs these EHRs when held by covered entities, which are health care providers, insurers, and information exchanges. The HIPAA regulations set out standards for security and privacy (HIPAA 2020). HIPAA distinguishes between individually identifiable information and information that has been de-identified to prescribed standards, in which case regulatory requirements no longer apply except to assure that the information will not be re-identified. Under HIPAA, many uses and disclosures of identifiable information require individual agreement, called "authorization," including uses for research or commercial purposes. Other uses and disclosures do not require any individual notice or authorization, such as uses or disclosures for treatment, payment, and health care operations, and uses or disclosures for public health as provided by law.

Thus HIPAA permits state or federal public health authorities to collect or use EHR information for public health without notice to individuals or their authorization. Health care providers may analyze data from their EHRs to assess community health needs or to decide which services to expand or close, also without individual notice or authorization, as this is considered part of health care operations. Public health researchers may not use EHR data that identifies patients without their authorization, however, unless the research qualifies for a waiver as explained below.

HIPAA gives individuals a right to request an electronic copy of most of the treatment-related information in their EHRs (called the "designated record set") or to request that a copy be transferred. Once EHR information has been transferred out by the covered entity, it is no longer within the protections of HIPAA. Likewise, the vast array of actors collecting health information in electronic form that are not

HIPAA-covered entities also are not subject to the HIPAA constraints. Instead, these entities are far more lightly regulated by the Federal Trade Commission Act's prohibition of unfair or deceptive trade practices. These entities collecting health information directly from individuals—including everything from Facebook to medical information sites such as WebMD to direct to consumer genetic testing companies—may have far greater scope of action, even when the information they receive originated from EHRs.

Within U.S., state laws may vary in the extent to which they protect information that has come into the hands of public health. These variations may create uncertainty on the part of the public about whether information transferred from EHRs to public health will be fully protected. One state law case has dealt with this issue. Journalists sought data in the state cancer registry obtained from patient records to investigate whether toxic exposures might have been related to cancer clusters of neuroblastoma in their region (Menderski ccc). They requested the information by zip code, date of diagnosis, and cancer type, but did not request any other identifying information. The health department refused to reveal the information, citing an analysis by Latanya Sweeney that patients could be re-identified through linkages with other publicly available data sets. The Illinois courts held that the Illinois Freedom of Information Act (FOIA) law permitted disclosure (*Southern Illinoisan* 2006). The trial court reasoned that the health department had not shown that the information would reasonably tend to reveal the identity of persons in the cancer registry—the statutory standard—because of the complex data skills required for the reidentification. The Illinois appellate courts upheld the ruling as correctly applying the state's statutory standard regarding information in the cancer registry. We discuss registries more fully below; suffice it to say here that without sufficient transparency about public health data collection from EHRs, people may be surprised to know what data the government possesses and how it may be used.

To take another example, in Utah information from a state data base of controlled substance prescriptions was released without a warrant as part of an investigation of drug diversion from ambulances. Two firefighters were erroneously accused of drug fraud based on information from the data base. The result was that the legislature passed a statute protecting the data base against warrantless searches by state and local law enforcement, but the federal Drug Enforcement Agency still has the ability to access the data base without a warrant (Whitehurst 2017). State auditors have also warned that cybersecurity protection for the data base is inadequate (Carlisle 2019).

Electronic health records (EHRs) contain vast amounts of patient data that might be used to supplement or substitute for at least some public health data collection needs. Although EHRs are now ubiquitous in U.S. health care, their use in public health has fallen far short of what many believe it could be (Friedman et al. 2013). The HIPAA privacy and security rules were something of an afterthought when concerns were raised that electronic transfers for insurance billing might threaten data privacy or security. EHRs are limited in the information they contain in part due to this history. Data are structured with fields that are relevant for billing and may contain other important information in natural language text that is more difficult to search (Hoffman 2018). They will only contain information as gathered through medical care encounters, including visits and information entered through portals where patients can enter their own information remotely. Information in EHRs will be incomplete if patients do not follow up after an initial visit or if they are seen by providers who use different EHR systems. There are also concerns about inaccuracies in the data entered into EHRs, although as physicians become more accustomed to technology the error rate may be decreasing (Hoffman 2018).

The fragmentation of health care delivery in the U.S. is another part of why EHRs are not as useful for public health as they might otherwise be. Different EHRs are used by different health care systems and the data in them cannot readily be analyzed together. Health information exchanges may be helpful in addressing this problem, as they process all data from providers and patients who participate (Saks et al. 2020). All payer claims data bases (ACPDs) are other efforts by state health departments to gather data about particular patients from all care for which payment claims were submitted. ACPDs have been very useful in developing fuller pictures of costs of care and of care utilization. For example, they can trace whether reduced outpatient costs of care for diabetic patients are correlated with more frequent hospitalization, which would raise costs and signal inadequate blood sugar control, or whether the reduced costs may be due to improved patient management. But ACPD data contain the information needed for reimbursement rather than a fuller picture of the patient's condition. Moreover, a Supreme Court decision in 2016 held that the federal Employee Retirement Income Security Act (ERISA) preempts state requirements for employer health insurance plans to submit their claims data (*Gobeille* 2016). This decision may eventually mean that fewer plans submit data to ACPDs and likely has discouraged the development of ACPDs in more than the sixteen states that currently have established them (Curfman 2017).

Public health itself is fragmented, too; responsibilities for public health lie primarily at the state level and may be devolved to local health departments which lack extensive information technology infrastructure (Williams and Shah 2016). EHR utilization in the US was heavily incentivized as part of the economic stimulus response to the 2008 recession. Providers who met standards for "meaningful use" of these records received extra payments and EHR use was eventually required for Medicare reimbursement. The actual public health benefits of the meaningful use incentives were limited because some public health systems were unable to receive data transfers from EHRs. Meaningful use for public health as it was implemented after the 2008 recession did include the ability of EHRs to interface with immunization and cancer registries, the ability to report cases to other specified registries, and the capacity to retrieve and submit data for syndromic surveillance, primarily as optional functionalities. It also included the capacity to retrieve and submit certain reportable clinical laboratory results such as positive tests for hepatitis. These reporting capabilities are only a small subset of the legal requirements imposed on health care providers to report patient information to public health, however. Moreover, unless EHRs are standardized in how they record aspects of disease and patient function, comparisons will continue to be elusive and EHRs will not contribute critical information about population health trends (Kruse et al. 2018).

Information equity is another significant problem for EHRs. Minority popula-tions in the U.S. receive health care less frequently and less adequately than other population subgroups. EHR data are likely to be better in quality and more complete for patients with consistent insurance coverage (Saks et al. 2020). Notably in regard to information equity, non-white and ethnic minority populations express privacy concerns about the use of EHRs at far higher levels than do U.S. whites (Clayton et al. 2018). The result may be that these populations are less likely to agree to use of their data, including uses that may benefit population health or subgroups of which they are members (Saks et al. 2020).

5.5.2 EHRs in the UK

Because with the National Health Service (NHS) health care in the UK is far less fragmented than in the U.S., the UK presents greater opportunities for use of patient care data for public health. Challenges remain, however. Among them are that in practice there is fragmentation of information technology systems across the UK and data sharing agreements are complex (digitalhealth 2016). Public concerns about privacy risks and the importance of ensuring patient trust in data use have come increasingly to the fore. A series of reports under the leadership of Dame Fiona Caldicott has addressed these tensions in data use through the development of what have been named "Caldicott principles."

The original Caldicott principles were:

(1) every use or transfer of confidential personal data should be clearly defined, scrutinized, and documented; continuing reviews should be regularly reviewed; these should all be conducted by an appropriate data guardian.
(2) Identifiable information should be only used when essential for the specified purpose.
(3) Identifiable data used should be the minimum necessary.
(4) Access to personal confidential data should be on a strict need to know basis, and then only for the minimally necessary data.
(5) All those handling personal confidential data should be fully aware of their obligations to protect confidentiality.
(6) All use of personal confidential data should be in accord with law.

A seventh Caldicott principle, adopted in 2013, is that the duty to share information can be as important as the duty to protect patient confidentiality.

The NHS Constitution stipulates that patients have the right to request that their confidential information not be used beyond their own care and treatment, to have their objections to additional uses considered, and to be given a legal explanation where their wishes cannot be honored (Caldicott 3, 2016, at 5). This constitution applies to health care but not to social care such as home help with activities of daily living. Controversy about NHS data use emerged in 2014 as the NHS sought to make greater use of technology to improve patient care. Patients were told by the NHS that records from their general practitioner would not be shared in identifiable form with the Health and Social Care Information Centre (HSCIC) if they so requested. They were also told that they may request that the HSCIC not further share their identifiable information. The NHS pamphlet with this information was confusing and resulted in the request for Caldicott 3, which produced a third version of the Caldicott principles.

Caldicott 3 relied importantly on the UK National Data Guardian (NDG) review of data protection to safeguard patients (Caldicott 3, 2016). It proposed a system under which patients could opt out of use of their records, rather than having to opt in to record use. This approach significantly increases the likelihood of record availability, since many patients are unlikely to take the steps needed to opt out. Patients were to be given two opt outs; declining to allow their records to be used to improve local services and running the NHS, and declining to allow their records to be used for research and improved treatment—although this last proposal was to be put out to further study. The basic value underlying Caldicott 3 was the need to enhance trust despite allowing record use. The Report begins: "This is a report about trust." (Caldicott 3, 2016, at 3) Important aspects of Caldicott 3 are recommendations for clear descriptions to patients about when information might be used, the requirement that there should be no surprise that an appropriate professional has access to information, and patient confidence that only the minimum necessary information is shared. Other aspects singled out as critical to trust included adequate data security, full public consultation, and dialogue with the public. The NDG could allow anonymized data to be used without consent, but subject to enhanced sanctions for failures to protect this anonymity.

The British standard for privacy rests in "reasonable expectations" in the sense that there should be no surprises to an individual about how data are used (Office for the National Data Guardian & Connected Health Cities 2018). To understand the idea of reasonable expectations (and whether it differs from the idea that there are circumstances in which consent might be implied), the NDG convened a three-day citizens' jury in 2018 (NDG 2016). The jury was given scenarios about a hypothetical patient and asked to discuss them in light of the legal standard for reasonable expectations: what an "average person of normal sensibilities" would think about whether the information sharing would be surprising and unacceptable. Jury answers thought information sharing among providers to improve individual patient care would be expected. They were less certain about the need to share information with administrators and other non-clinical personnel beyond information that would be minimally necessary to contact patients, schedule appointments, and bill for services. Expected sharing included discussions with experts at other facilities about difficult care questions. Jury members were not as sure, however, that patients would reasonably expect information to be shared to help experts diagnose their own patients. A majority of the jury did not expect information to be shared with patients' relatives; they thought that without consent this sharing would not improve care and might lessen trust. Extended discussion of reasons for and against data sharing increased the likelihood that jurors would want to share the data.

In 2018, the National Data Guardian (NDG) was established in the UK as a statutory office, enabling the NDG to issue official guidance about processing of health and adult social care data. Dame Fiona Caldicott was appointed as the initial holder of the position in early 2019. Commentators have argued that with this arrangement

Britain could be a model for balancing use of data to improve health in an aging population with protecting privacy (Chan et al. 2016).

5.5.3   EHRs in the European Union: The General Data Protection Regulation and Public Health

The European Union's General Data Protection Regulation (GDPR) went into effect in May 2018. It sets out a common set of regulations that all EU members must adopt for data protection. EU member states are in the ongoing process of statutory revisions to implement the GDPR (Molnár-Gábor 2018, on Germany). According to Molnár-Gábor, European privacy regulation is most strongly influenced by French and German law; the fundamental anchor of French law is personal integrity and the fundamental anchor of German law is human dignity and the right to informational self-determination.

The advent of the GDPR caused considerable consternation about the availability of sensitive health information for research or public health, although there are exceptions to the requirement of individual consent for both. Genetic and biometric data are specifically mentioned as sensitive. Explicit consent of the data subject is required but there is an exemption for research when there are suitable safeguards and the processing is carried out for reasons of substantial public interest. Suitable safeguards may involve an oversight body such as an ethics review committee (see description in Shabani and Borry 2018). Pseudonymized data is treated as within the exemption for research. (This is data where a key code, kept separately, is substituted for identifying information. Records with the same key code can be linked, but only those with access to the key will know who the individuals are).

For public health, Recital 54 of the GDPR provides that the processing of special categories of personal data may be necessary for reasons of public interest in the area of public health without consent of the data subject. When this is done, the processing must be subject to suitable, specific measures to protect the rights and freedoms of natural persons. Public health is interpreted as in Regulation (EC) no 1338/2008 to involve all elements related to health, including health status, mortality and morbidity, provision of health care, access to health care, health needs and resources, health care expenditures and financing, and causes of mortality. Recital 54 stipulates that data processing for public health should not result in processing by third parties such as employers, insurers, or banking companies for other purposes (EU 2019).

Researchers and public health advocates voiced concerns that the GDPR would discourage EHR use for these purposes despite the stated exceptions. COVID-19 and the need for rapid collaboration about disease spread brought these concerns to the fore (McLennan et al. 2020). Digital patient records, if available across systems and countries, are a critical source of information about the disease incidence, course, and possible treatments. Many of the early research about COVID-19 relied instead on small case series of patients in a single institution and thus did not provide evidence regarded as high quality. Because of differences between the U.S. and the EU regarding data protection, proposals for combining data sets across the Atlantic may be legally difficult. McLennan and colleagues, however, argue that solidarity requires pursuing this option and that the GDPR does permit data use for this purpose with appropriate technical and organizational safeguards.

The EU has embarked on an initiative on the Digital Transformation of Health and Care (Digicare). There is a proposal to create an innovation cloud within the Digicare environment that would be consistent with GDPR protections (Aerestrup et al. 2020). One of the goals of this technology is to be able to utilize shared computational and storage resources that can enable processing of huge data sets needed to detect weak signals about possible correlations. Another is to enable use of pseudonymized data that cannot be downloaded so that individual identities can be protected. As these and other initiatives continue to develop, transparency, oversight, and purpose specification will continue to be important to prevent mistrust. It seems likely that people will be far more willing to accept the use of data sets such as these to provide early warning signs of public health emergencies or to gain urgently needed knowledge such as for COVID-19 than for purposes that are not as immediately urgent.

5.6   Bloodspots Retained from Newborn Screening

Since the 1960s, newborns with access to good health care have been screened for metabolic conditions such as phenylketonuria (PKU) that require early treatment to prevent severe disability or even death. Screening programs represent a population-based intervention with a very strong public health aim: identifying individual health conditions that can be addressed through timely interventions. Although the conditions now screened for may vary with the jurisdiction, they typically include sickle cell, cystic fibrosis, PKU, congenital hypothyroidism, and certain other inherited metabolic conditions. The screening is conducted by a heel stick; blood spots are collected on a card. If they are retained, the blood spot cards can be an enormously rich source of material not only for research about genetics but also for public health issues such as infection seroprevalence, toxic exposures, epigenetics, and other information available from a blood sample. The samples have the potential to form a population-wide data bank of genetic information and environmental information available from blood that could be linked to clinical outcomes on an ongoing basis (Lewis et al. 2012). The information in the samples that is genetic can, with sufficient other information, be linked back to individuals and thus is not fully de-identifiable. It could in principle provide important warnings to individuals about carrier status or risks of adult onset conditions.

The information obtained from retained newborn screening samples has several features that are especially important ethically. It is collected from individuals (newborns), who have no knowledge of the collection and who are unlikely to ever become aware of it unless they are told about it, or encounter consequences from it, in later life. In many jurisdictions, the screening is mandatory, although retention of the spots beyond the short time periods needed for confirmation and perhaps also program improvement may require separate consent. If consent is required, it is given by parents on behalf of their children; such consent, however, comes at a time of great excitement when attention is likely to be elsewhere. Parents may remember the heel stick if the baby cries, but memories that bloodspots may be saved for future research or public health are likely to fade very quickly. Because newborn screening applies to everyone, retained samples perhaps also represent the most comprehensive snapshot of the population at any given point in time; these spots are the available data least likely to omit vulnerable population sub-groups, at least in jurisdictions where screening is mandatory. They also will contain complete information from age cohort to age cohort, so they present the opportunity to make comparisons among individuals born in different places and times. Just knowledge of the date and place of birth about an individual, however, may convey critical, powerful, and potentially stigmatizing information about them, depending on what has been learned from these samples. Finally, the genetic information in them can be linked to individuals, although the samples may also be useful for studies such as levels of environmental exposures at a given point in time that do not require individually identifying information.

A comprehensive study of the use of retained newborn screening samples in Denmark illustrates the potential uses of this resource (Nordfalk and Ekstrøm 2019). Newborn screening and bloodspot retention for research are routine in Denmark. Although the legal possibility to opt out of the screening exists, no informed consent is required for sample storage and use in research. No records are kept of the frequency with which samples are actually used in research. Nordfalk and Ekstrøm used a survey of publications to identify secondary uses of screening samples and their frequency. Their survey strategy was the best available but did not capture any uses of the samples that were not published. The researchers estimated that just under 40% of the samples had been used for purposes other than newborn screening. Most but not all of the studies had research ethics approval. The most-studied disease topic was mental illness (about one-fifth of the uses), followed by metabolic disorders and diabetes. It seems possible to hypothesize that individuals would be less likely to expect newborn screening samples to be used in studies of mental illness than in studies of diseases of a type more closely resembling the conditions represented in the screening panel itself, which are largely metabolic disorders. Nordfalk and Ekstrøm argue that information about actual uses of samples is important for both discussion of the permissible range of uses and for public transparency and trust.

For many years, blood spots obtained in screening were routinely retained by programs in many U.S. states and other jurisdictions. Families—and of course the newborns themselves—were utterly unaware that the samples could be made available for a variety of research, public health, and other uses. When awareness of the existence and use of the samples emerged, it sparked protests and litigation by privacy advocates in Texas, in Minnesota, in Ireland, and elsewhere. The result of litigation in Texas and Minnesota was destruction of millions of samples. Some have judged the destruction a disaster for public health. At best, the destruction is a clear example of what can happen when transparency in data use is lacking and people lose trust in data retention (IOM 2010).

Concerns about blood spot retention have adversely affected newborn screening—and thus health, too. While in most jurisdictions screening is considered routine, some jurisdictions permit parents to opt out. Some even require explicit parental consent for the procedure. Although by far the majority of parents are supportive of the method for identifying the low risk of serious disease, some object to the pain of the needle stick for their child, some have religious objections, and some find methods of communicating positive results unsatisfactory (Etchegary et al. 2016). A great concern of public health officials is that repurposing of screening samples will scare people away from allowing their children to be screened—and that even the knowledge that such uses may be possible will be sufficient disincentive for screening (Editorial 2011). To avoid risks that parents will opt out, the U.S. National Society of Genetic Counselors supports mandatory screening and uniform, transparent policies to govern any use of residual samples (Blount et al. 2014).

Jurisdictions such as Michigan now have moved to separate blood spot retention from initial screening by adopting opt-in consent mechanisms to govern retention of blood spots for research or public health. The Michigan BioTrust retains samples for up to 100 years. Use of the samples for research requires parents to opt into this possibility shortly after delivery. The Michigan BioTrust also contains legacy samples obtained between 1984 and 2010, before initiation of the opt-in process; these samples may be used in research unless individuals request their destruction (Michigan DHHS 2019). Parental opt-in rates to research use are low, possibly because parents are never asked to opt into this use during the period of excitement and confusion surrounding birth of a newborn. Pilot programs using social media to enhance awareness have not been particularly effective, although further development of these mechanisms might enhance transparency and understanding about ongoing collection and retention practices (Platt et al. 2013). Michigan's experience suggests that without more discussion and changes in practice, opt-in consent is unlikely to yield blood spot retention rates that are useful for public health.

In general, commentators remain concerned that participants and the public are largely unaware of the possibility that blood spots might be stored and re-used. Some uses may be judged unacceptable by many (e.g. Thiel et al. 2014). A study of public preferences in Canada also indicates that people prefer that permission be asked and that discussion that increases understanding of the important research uses of the information for health may increase willingness to allow samples to be used without permission (Hayeems et al. 2016). A study in Illinois revealed similar preferences for permission, along with demographic differences in support. Willingness to support research use was more likely to be found among those with higher levels of education and less likely to be found among Blacks (Hart et al. 2015). These findings suggest the importance of reasonable expectations and the concerns that might arise with surprises about data use. They suggest that if uses of retained samples are not transparent and people are surprised about data collection and use, protests like those that occurred in Minnesota and Texas may recur. Finally, they suggest continuing concerns about discrimination in research and health care in the United States.

Uses of retained blood spots to increase understanding of disease and improve population health may also result in information that can be useful for the health of individuals. For example, it may indicate that individuals are at highly elevated risks of developing certain cancers, Huntington's disease, or dementias. Whether, when, and how, any such results should be returned to individuals raises additional vexing questions. On the one hand, information return may diminish trust, especially if the information is troubling, raises questions about identity, carries risks of discrimination, or creates anxiety and uncertainty. Imagine learning suddenly about risks for a serious disease, without forewarning and without any previous knowledge that a governmental entity had information about the risk. On the other hand, the fact that some have potentially life- or health-altering information about others, but do not reveal it when it could be beneficial or actionable in some way, is also troubling. It may be especially troubling if the institutions holding the information are public health authorities—authorities that have improving health as their mission. Moreover, if such failures to reveal critical health information come to light, they may cast added suspicion on the function of newborn screening programs. A further complexity is that if the secondary use of the samples occurred without awareness or any kind of agreement, no information will be available about individual preferences for return of results. These preferences may be especially important for adult-onset conditions and for conditions that are not clinically actionable, such as Huntington's disease, but that nonetheless may be of great importance to non-health related choices such as whether to have children or save for retirement (Lewis and Goldenberg 2015).

Bloodspot retention presents one of the clearest illustrations of the need for communication and transparency. Yet the timing of sample collection remains challenging: samples are collected at birth, any consent is not obtained from the individual from whom the sample is taken but from parents, and sample utility may not become apparent until far in the future. A sense of surprise when sample use came to light generated concern, protest, destruction, and regulation. Not only was public health the loser, but in some jurisdictions, newborns also were put at risk when their parents refused the testing.

5.7   Biobanks

Recognition of complex relationships between genotypes and phenotypes and of the importance of genetic information in tailoring treatments to individuals has led to extensive efforts to collect data bases of genetic information and banks of biological samples, especially when these can be linked to clinical records on a continuing basis. Some of these biobanks are established by entities administering health care to individuals. For example, Intermountain Healthcare has been collecting a biobank of tissue samples and patient outcomes that can be used for precision medical care for patients with cancer and other conditions (Intermountain 2019).

Other biobanks are society wide. For example, UK Biobank is a charity that recruited 500,000 people from across the UK beginning in 2006; participants gave blood samples and agreed to have their medical information collected on a continuing basis. On entering the UK Biobank, participants gave what is called "broad consent": consent to the ongoing collection and use of the data for "health-related" research (UK Biobank 2018). Patients who opt out of use of their NHS records for research pursuant to Caldicott 3 do not thereby opt out of UK Biobank; that would require separate notice to the biobank. All data made available for research by UK Biobank are de-identified; the biobank maintains identifying information only to properly establish linkages.

Denmark has one of the most extensive biobanking syvstem. The Danish Biobank Register lists the Danish National Biobank, which includes newborn blood spots since 1982; the Danish twin registry with data about 86,000 pairs; the Danish Cancer Biobank; and information from many other biobanks in Denmark. All together, these banks contain information about 5.7 million people, nearly the entire population of Denmark; they can be linked to other data such as location of residence that may be especially important for health (Staten Serum Institut 2020). Commentators have raised concerns that members of the Danish public are largely unaware of the numbers of studies that have used the Danish National Biobank (Nordfalk and Ekstrøm 2019). All studies using samples in the Biobank must be approved by the Danish Data Protection Agency, which can waive consent if there are no health-related risks and the use does not burden the participants. Some commentators have questioned whether the GDPR exceptions for research provide adequate protection in the case of biobanks, and have urged that national laws implement oversight and transparency standards (Staunton et al. 2019).

5.8 Registries

Registries collect information from or about people who share a common procedure or condition. The information they contain may be gleaned from medical records, from research, or directly from patients or their families. Some registries are maintained by public health agencies, some by practitioners or medical societies such as cardiology, some by researchers, some by charitable organizations devoted to particular diseases, and some even by pharmaceutical companies (Francis and Squires 2018). Registries may be international in scope, especially those devoted to rare diseases. Some registries only contain information as minimal as a list of contact information for people with a given condition who may be interested in participating in research about the condition. Others follow people throughout their lives, assessing disease course, treatment outcomes, and quality of care. People may be entered into registries as infants or children when they are first diagnosed, but in the United States will be asked to re-consent to participation when they reach adulthood (Francis 2014).

These registries may serve as valuable surveillance methods for assessing potential causes of increases or decreases in disease occurrence. For example, in the U.S. under the Surveillance, Epidemiology, and End Results (SEER) program of the National Cancer Institute, every state maintains a tumor registry (NCI 2021). These registries collect information about all cases of cancer diagnosed in the U.S., including identifying patient information, tumor site and morphology, tumor stage, and initial treatment. Deidentified information from these registries is released to the SEER program to compile population statistics about cancer in the U.S. It was supposedly deidentified information that was released from the Illinois cancer registry that was the subject of the controversy described earlier in this chapter.

According to Fairchild and colleagues (2007), when the U.S. adopted the Privacy Act protecting information held by agencies of the federal government, it did not include a proposal for people to be notified when information about them was collected by the federal government. U.S. law governing the confidentiality of medical records does not require that patients be notified when information in medical records is shared with public health (HIPAA 2020). Patients are not made aware that data drawn from their medical records has been shared with the SEER cancer surveillance program, for example. Law professor Wendy Mariner (2007) has argued that these means for sharing information from patients' health records with public health without consent are violations of the constitutional right to privacy. Others argue that the statutes are sufficiently narrowly tailored to address important state interests in protecting health (McLaughlin et al. 2010). We think it might even be argued that people who are receiving care from shared resources should reasonably expect that data from them may be used on an ongoing basis to improve health—but only if the considerations we outlined at the beginning of this volume such as transparency, protection from harm, and equity are appropriately implemented.

In the U.S., the non-profit Cystic Fibrosis Foundation maintains a registry of people with CF who agree and are treated in centers accredited by the Foundation. The registry supports research about CF care, uses data to improve care quality, and follows the natural history of the disease and care outcomes (Schechter et al. 2014). Similar registries exist elsewhere as well, such as the UK CF Registry. Drug companies also sponsor or support registries, sometimes in cooperation with disease foundations, in the effort to develop treatments that may prove effective.

There are also registries organized on an international basis to collect information from individuals with rare diseases. For example, the Orphan Disease Center has built registries that seek information directly from patients, including information from their medical records (ODC 2020). Although these international efforts may be critical for the understanding of rare conditions, their efforts to collect data also raise complex questions about meeting the legal and ethical requirements of different jurisdictions for data protection. Given the proliferation and variety of registries, it is not surprising that people might be confused about the entities that possess and use health information for many different purposes.

5.9   Information Gained in Medical Research

Information gained for or in the conduct of medical research might also be a useful source of information for public health. The many research biobanks are a good illustration of information that can be used to link individual genotypes and outcomes. The U.S. "all of us" precision medicine initiative to construct a biobank for a million lives is designed to be one of the largest and most representative of these, but there are many others. For example, the U.S. National Library of Medicine maintains the database of Genotypes and Phenotypes (dbGaP) to collect the data and results of genetics studies in humans funded by the federal government (NCBI 2019).

When patients agree to participate in medical research, they give explicit informed consent to the project. This informed consent typically includes information about who may have access to the information gained in the research, how the information will be protected, and how the information will be used. The consent may be very broad—for example, to any further use in research studies—but it is consent nonetheless. Thus research participants may have particular grounds for surprise if data gained in the research are used for new purposes. This kind of surprise is envisioned by the Fair Information Practice principle that individual consent should be sought for secondary uses of data. Even when people are willing to allow data originally collected as part of a research project to be shared, one study indicates, they still want to be asked (Ludman et al. 2010).

These features of medical research suggest that the case against repurposing of research data for public health is stronger than the case against repurposing data from health care. Reactions may be particularly intense when consent to the research was for a topic of particular health interest to the participants, when the public health purpose is far different, or when the use appears inequitable or stigmatizing. A further concern is that consent was sought initially and may not have included any further data use. The highly publicized case of genetic research involving the Havasupai Indian tribe illustrates. Researchers from Arizona State University conducted research on diabetes using blood samples from tribal members. Members of the tribe were encouraged to participate in the trial because of the particularly heavy health burden of diabetes for them. When researchers later used the samples that had been stripped of identifiers to study schizophrenia, migration patterns, and tribal inbreeding, the tribe were outraged. The tribe sued, and the University ultimately settled for compensation and destruction of the samples (Sterling 2011). This of course was a second research study, not a study by public health. Notably, it also used data that were considered deidentified. But it is not difficult to imagine similar outrage if the state public health department had used the data to assess whether tribes were experiencing high burdens of schizophrenia or inbreeding. Finally, given the specific inclusion criteria for many research studies and the widespread availability of other data, it seems unlikely that medical research data will be uniquely important as a source for public health. In most circumstances, therefore, the balance of considerations would seem to weigh against repurposing data from medical research for public health. At a very minimum, however, if it does occur, it should be subject to the same constraints as the use of information from health care.

5.10   Direct to Consumer Testing, Including Genetic Testing

Direct to consumer testing is another potential resource for public health. Consumers now submit samples to commercial companies such as 23andMe or Ancestry.com for personal enjoyment and in some cases health testing. These companies offer consumers information about their ancestral origins and possible relatives. They also may provide lower-cost alternatives for identifying deleterious variants of genes such as BRCA1 or BRCA2. For example, 23andMe now offers an approved test for three of the most common variants of these genes that are associated with high risks of breast cancer (23andMe 2019).

When individuals sign up for these direct to consumer products, they are given terms of service and asked to consent to uses of the data. Companies vary widely in the protections they provide to consumers (e.g. Bunnik et al. 2014). Jurisdictions also vary widely in the legal protections that they give consumers. For example, the U.S. the primary protection at the federal level is the Federal Trade Commission Act (15 U.S.C. § 45 (2019)) prohibition of unfair or deceptive trade practices. This prohibits companies from misrepresenting their data protection practices but does not prohibit companies from explicitly declaring that they will be quite open about sharing information they have gained. States also have their own protections of unfair trade practices that may mirror the federal.

Controversy has erupted in the U.S. about whether these data bases may be used for law enforcement purposes. For example, GEDmatch is an online service that allows individuals to upload data that they have obtained from other sources such as the direct to consumer testing companies. Individuals who upload their data may search for other relatives who have submitted their DNA. Law enforcement accessed GEDmatch to find relatives who might be linked to a killer who had been responsible for multiple murders, rapes, and burglaries in California between 1974 and 1986, and whose identity had frustrated law enforcement. This successful search for the "Golden State killer" highlighted the potential utility of these data bases to solve crimes but also the privacy risks to people who had uploaded data without imagining its crime-fighting possibilities (Kaiser 2018). It is not hard to see that public health might have similar interests in this information, for example if genetic material is available about an unknown individual who has been the source of a deadly infection.

Preliminary studies indicate public support for access to these databases to solve violent crimes, crimes against children, or sexual assaults, but far less support than when the access is to solve nonviolent crimes (Guerrini et al. 2018). This would suggest that there might be support for public health to consult these data bases in the unlikely event of the immediate need to identify someone who is at imminent risk of transmitting deadly infections, but not otherwise. If public health were permitted to do this, moreover, it should be subject to the ethical concerns we raised about contact tracing in Chapter 2. With these constraints, public health access would at least minimize additional risks to persons thus identified.

5.11 Smartphones and Smartphone Apps

According to estimates by researchers at the Pew Research Center, over 5 billion people used mobile phones in 2019; this represents just over 70% of the world's population. About half of these devices are smartphones. These percentages are higher in advanced economies and more variable but growing rapidly in emerging economies. Smartphone ownership rates are higher among people who are better off, better educated and younger in age. Even in some advanced economies, younger adults are far more likely to own smart phones than older adults; in Japan, for example, nearly all adults under 35 (96%) own smart phones while far fewer adults over 50 have these devices (44%). To illustrate with one notable gap by education, in Nigeria 51% of people with a secondary education or more use smart phones, whereas only 6% with less education do so. In most emerging economies, men and women use smartphones at roughly comparable rates; the exception is India, where men (at 34%) are over twice as likely than women (at 15%) to own smartphones. Given these differences, data collected from smartphones will have notable gaps, particularly among the elderly, the less well educated, and the poor. These gaps are closing in more advanced economies but at least for now are growing in less advanced areas as smartphone adoption is rising vary rapidly among the young and better off (Taylor and Silver 2019).

Despite the gaps, smartphones are judged to be a significant advance in their potential for disease surveillance. Mobile phones allow information to be transmitted quickly, cheaply, and from locations where communication may otherwise be unavailable. Smartphones add information, too, especially about location in real time (e.g. Lee et al. 2016). Efforts to address malaria transmission have taken particular advantage of cellphones. The World Health Organization used smartphone technology to assist in the efforts to eliminate malaria in Bhutan. The technology allowed essential real-time transmission of geolocation information (WHO 2019). A group of researchers have used travel surveys and cellphone data to provide a more complete map of transmission of malaria in Bangladesh, including travel from urban residents to remote rural locations (Chang et al. 2019). Without the analysis of the mobile phone data, it was widely believed that travel to the countryside was infrequent and did not play a significant role in transmission; it is now understood that people travel home fairly often. Still, gaps in the data remain, as in rural areas cellphone towers are less frequent. An earlier study by some of the same researchers used mobile phone call data records to assess the travel patterns of 15 million cell phone owners in Kenya, allowing them to create detailed risk maps of the movement of malaria parasites (Wesolowski et al. 2012). A different group of researchers studying public health surveillance using mobile phones in sub-Saharan Africa concluded that the primary use of cellphones in surveillance was enabling public health workers to transmit timely information about infectious diseases such as malaria or influenza (Brinkel et al. 2014). The transmissions often were done by public health workers using their own phones, raising questions about the source of repayment for the phone use. A major advantage of cell phone use, however, was that it allowed for two-way communication. This study concluded that cell phone use for surveillance in sub-Saharan Africa remained fragmented and small in scale. Cellphone data also were used during the Ebola epidemic to assess the need to control movement and the efficacy of travel restrictions (Peak et al. 2018).

Twitter, one of the most frequently used smartphone apps, also provides geolocation information. One group of researchers used information from Twitter to understand the triggers of influenza epidemics. These researchers found that the strongest predictors of influenza incidence were the host populations socio- and ethno-demographic properties, specific weather traditions, viral antigenic drift, and the host population's land travel habits, among others. To reach these conclusions, they used multiple data sources including Twitter: insurance records capturing the dynamics of influenza-like illness, weather measurements, air travel and geographic proximity, Twitter data revealing long and short distance human movement patterns, and census data (Chattopadhyay et al. 2018).

In addition to the enhanced communication they provide, geolocation data in real time and the ability to track movement over time are at present the primary additional contributions of smartphone data to surveillance. As the research we have described indicates, these contributions may be highly informative additions to the data that can be accessed by public health from other sources. However, these capabilities enhance risks to individuals because they enabled them to be located and tracked in a way that information in static databases does not. These risks increase the need to ensure that, if these methods are used, they are coupled with the protections that we discussed in Chapter 2 for advanced COVID-19 surveillance.

In addition, geolocation information from smartphones enables individuals to be tracked over extended periods of time. In the words of the U.S. Supreme Court, "A cell phone faithfully follows its owner beyond public thoroughfares and into private residences, doctor's offices, political headquarters, and other potentially revealing locales ….Moreover, the retrospective quality of the data here gives police access to a category of information otherwise unknowable" (*Carpenter* 2018, p. 13). If cellphone information is routinely used to monitor individuals' movements across time and space, individuals may come to fear that a shadowy Big Brother really is watching over them. Many commentators have raised the alarm that continual surveillance will chill freedom of thought and expression, especially when people are tracked over time by what they perceive to be agents of the government (Gullo 2016). These concerns indicate the importance of oversight, transparency, and limited use by public health of devices that can track movement over time.

5.12   Robots, Wearables, and Biosensors

Another potential new data frontier for public health is the presence of devices that can observe the individual directly. Robots may watch an individual's every movement, recording everything the individual says and does. Wearables track exercise, sleep patterns, blood pressure, and blood sugar levels. Biosensors may identify whether individuals have taken medication, how that medication has been absorbed, and even features of the individual microbiome.

Domestic robots are common in many households in wealthier countries. "Companion" robots are advertised as a way to make it easier and safer for older people to live alone and remain in their own homes. "Buddy," headquartered in France, advertises that it can provide companionable social interaction, medication reminders, easy access to the internet, and warnings about falls for seniors who would otherwise be isolated (Blue Frog Robotics 2019). But Buddy could just as easily send messages to public health about the health status of the elderly in their community. Care robots are increasingly used in care facilities to aid in feeding, hygiene, and monitoring; designed to look like pets, they can also provide comfort and measure distress. These robots are cost effective and can extend personnel when there are labor shortages. But they are also ethically controversial. A recent systematic review of ethical arguments about care robots concludes that respect for privacy, behavioral control, and deception were major ethical concerns (Vandemeulebroucke et al. 2018). This review concludes that transparency and stakeholder engagement are critical to determining whether robot use is ethically permissible in particular circumstances. Like in-home robots, these robots could conceivably transmit a great deal of information to public health—from information about the moods and possibly quality of life of people in their care, to information about abuse or neglect in care centers.

Wearables also collect a great deal of health information, often through smartphone apps. Fitbit measures steps and sleep patterns, multiple apps are on the market for measuring blood pressure, and the iPhone will measure breathing patterns and heartrate to help users with mindful meditation. Many people use these mechanisms for personal enhancement, but they also can be recommended by health care providers and tethered to electronic medical records. For example, the British National Institute for Health and Clinical Excellence recommends wearables for blood glucose monitoring in certain circumstances. The devices can also be used to track periods of fertility for people who want to conceive—or to avoid conception. Whether these constant monitors do more harm than good was the subject of the medical ethics debate between Oxford University and Cambridge University in 2018. Although Oxford won the debate for the side in favor of their use, considerable concerns were raised about the risks of third-party access to data on cellphones left casually around, data security protection, and access to data by cellphone providers and app designers for their own purposes. Another major issue was access by the government to data stored on cellphones. Such constant monitoring of bodily functions may also medicalize everyday life, worrying the well and demanding hyper-vigilance on the part of the sick (Gilmartin et al. 2018).

Biosensors may be even more intrusive than wearables. In 2017, the U.S. Food and Drug Administration gave marketing approval for the antipsychotic Abilify MyCite in a form that allows tracking of whether the drug has been ingested (FDA 2017). It is not difficult to imagine applying the same technology to controlled substances such as opioids to assess whether they have been ingested by the patient for whom they were prescribed or been diverted. Such mechanisms could also be applied to whether women have taken their contraceptives, data of possible interest to boyfriends concerned about deception or Medicaid programs concerned about whether women are staying on prescribed birth control. Biosensors can measure dietary intake, alcohol consumption, or use of illegal drugs. The day is not too far off in which it will be possible to continuously monitor the entirety of the gut microbiome.

The potential for public health of these monitoring devices is not at all farfetched. There are reports of their use in China to ensure that people who have tested positive for COVID-19 remain in self-isolation. Moreover, they will yield information that cannot be acquired as readily in any other way. While we have presented these devices in a provocative manner, we have done so to reflect what may likely be public reactions to collection of data from them by public health. Any hint of such collection is likely to be met with revulsion and resistance. Reactions are likely to be especially strong if the collection comes from within the body. At present, we think, public health would be well advised to stay away from such biosensor measurements, except in very carefully defined circumstances in which robustly informed consent is obtained from willing participants.

5.13 Public Health Surveillance by Actors in the Private Sector

Many non-state actors are now engaging in what might be regarded as public health surveillance—that is, surveillance *of* public health even though not surveillance *by* public health. Many of these actors are using some of the novel forms of data that we have just described. These actors range from formally established partners with the World Health Organization, to health care providers, to internet search engines such as Google and social media sites such as Facebook.

Reliance on non-state actors for public health surveillance is not new. Amy Fairchild and colleagues describe how from colonial times innkeepers, ship operators, and even family members were obligated to report deadly contagion (Fairchild et al. 2007, p. 2). They locate the professionalization of public health—and the reservation of public health functions to public health—with the development of bacteriology in the late nineteenth century. We might also note that the late nineteenth century was a time of more general professionalization, not only of medicine but also of law, although medicine was a particularly powerful instance of the trend. Kristin Luker's study (1984) of the history of reproductive policy, for example, reveals how the efforts of physicians to take control from midwives changed policies around abortion and childbirth. That non-state actors might have resources and skills beyond what are available in the public sector is also not new. What might be newer today is the extent to which entities in the private sector are interested in surveilling for their own purposes along with public health and the scale on which they are so doing. And what is also newer are the kinds of data to which they have access and the power of the analytical tools they use.

Public health governmental agencies, moreover, are not the only entities engaging in activities that might be regarded as public health surveillance with new data forms. These private sector actors may not be scrutinized or regulated to the extent that government agencies are subject to oversight. In the U.S., in order to claim exemptions from federal income taxes, non-profit health care providers operating hospitals must perform community health needs assessments every three years and adopt implementation strategies to address the needs thus identified (IRS 2018). Data from electronic medical records are a frequent new source of data that may be used for public health purposes by non-profit health systems as well as by public health agencies. Direct-to-consumer genetic testing contains a wealth of health-related information about individuals. Credit card expenditures reveal health information from purchases and purchasing patterns; grocery store loyalty cards show dietary changes and indiscretions. Massive amounts of health information are collected, analyzed, and transmitted by internet search engines and through social media sites. This includes misinformation, too. Smart devices from robots to refrigerators are highly revealing about day to day activities.

In addition to the kinds of questions that raised by novel and large-scale data, additional issues of oversight are implicated when non-state actors engage in surveillance. In some jurisdictions, notably in the U.S., these actors may be subject to limited regulatory restrictions on what they do. At the U.S. federal level, for example the primary statute regulating non-state commercial actors who are not governed by HIPAA is the Federal Trade Commission Act prohibition on unfair and deceptive trade practices, 42 U.S.C. § 45. Non-state actors may have their own interests, consume resources in ways that duplicate or even perhaps undermine the efforts of public actors, present conflicts of interest, or perhaps even act in ways that are judged by others to be misconduct. To illustrate with a conflict before the advent of big data between a well-intentioned charity and public health, Fairchild and her colleagues (2007, p. 152) describe efforts of the National Foundation for Infantile Paralysis during the 1940s to get direct reports of polio cases. The Foundation argued that they needed the data to get proper services and treatment to patients; public health professionals were concerned about unnecessary duplication and lack of coordination and argued that the Foundation's efforts should be supportive only. The tension reportedly faded with the development of polio vaccines.

5.13.1   WHO and Non-state Actors

Established over seventy years ago, the World Health Organization has become an international body that can identify and respond to public health emergencies and that can provide information and support for global public health. WHO was created at a time when there were few other international organizations with a public health mission. But WHO has not been uniformly successful. The new International Health Regulations that entered into force in 2007 (2005) vastly expanded expectations for surveillance, allowing WHO to receive reports from non-state actors and to investigate potential health emergencies of international concern beyond the limited list of contagious diseases that had previously been within its purview. Acting under these regulations, WHO was criticized for overreacting to the outbreak of influenza in Mexico in 2009 and for reacting far too slowly to the Ebola outbreak in West Africa in 2014. Criticism of the handling of the Ebola epidemic led to extensive debate over the effectiveness of the organization in addressing epidemics and in coordinating strategies to address their containment and reduction.

The debate over the role and effectiveness of WHO had not diminished even by 2019: "The planet's premier health agency has announced drastic reforms. Critics say they aren't drastic enough" (Kupferschmidt 2019). Problems cited by critics included persistent shortfalls in funding and difficulties in coordination between WHO headquarters in Geneva and regional offices. Eigil Sørensen, a former WHO official who is now on the public health faculty at Thammasat University in Thailand, writes that national public health agencies, once dependent on WHO for technical assistance, have gained a great deal in competence although they continue to rely on WHO for its publications on global health statistics (Sørensen 2018). Over the years, Sørensen observes, organizations such as the Global Fund to Fight AIDS, Tuberculosis and Malaria, The Vaccine Alliance, and Unitaid have arisen in part out of frustration with what they judged to be the inadequacy of WHO efforts. But these non-state organizations may have their own priorities—priorities that may not always align with the priorities of WHO or with the priorities of other non-state actors.

The challenge for WHO is what role it may be able to play as new actors emerge in surveillance and other efforts to address public health. It may be able to play a mobilizing role in forming coalitions to respond to new pandemics or epidemics. Or, WHO may try to find new income streams that allow it to set budgetary priorities. Or, it might focus its role on serving as a forum to build consensus sustaining public health values, perhaps becoming an honest broker in disputes among member states. If non-state actors play increased roles in public health emergencies, as they are doing around the globe, unless WHO has adequate funding it may be in a position of stepping back into roles of advising and encouraging rather than serving as an active participant.

Under the International Health Regulations, the WHO is now permitted to take reports of potential health events of concern from non-state actors and to investigate these reports according to established science (WHO 2016, Art. 9). In 2016, WHO adopted a framework of engagement with non-state actors that recognized their importance in the promotion of global public health activities but also aimed to protect "WHO's integrity, reputation, and public health mandate" (WHO 2016, ¶ 4). This framework also included separate provisions for NGOs, commercial entities, philanthropic foundations, and academic institutions. In explaining the need for the framework, WHO singles out concerns for transparency, undue influence, conflicts of interests, and potential risks to its credibility raised by partnerships with non-state actors. WHO is also concerned that through the support they are given non-state actors may acquire competitive advantages, pursue their own interests, and whitewash their images. The WHO Director-General now reports annually on the status of engagement with non-state actors. The 2019 Report of the WHO director-general indicated both success with and ongoing challenges of establishing partnerships with non-State entities under the framework; challenges included operationalizing the requirement that non-State actors confirm that they do not further the interests of the tobacco industry and cosponsoring global health events with private sector entities with potential commercial interests (WHO director general 2019). It remains to be seen how the role of WHO in conjunction with non-state actors will continue to evolve.

If the US had followed through on withdrawing from the WHO, it would have been even more dependent on non-state actors for surveillance because of the funding loss.

5.13.2 U.S. Non-profit Hospitals and Community-Based Needs

In the U.S., non-profit hospitals are expected to engage in surveillance-like activities in order to qualify for exemptions from federal income taxes. These "community-based needs assessments" were introduced by the Affordable Care Act (ACA) and require the hospital to identify health needs of their communities and develop plans to address these needs as a condition to maintaining their tax exemptions. If the entire ACA is struck down by the Supreme Court as the Trump administration urged (Federal Respondents 2020), this statutory requirement would no longer stand, although hospitals would still need to demonstrate that they meet the standards of the Internal Revenue Service charitable tax exemption.

The needs assessment requirement is an opportunity for hospitals to coordinate with local health departments but also raised the concern that there may be inadequate consultation with public health in developing priorities that engage the public and are equitable (ASTHO 2012). A 2019 study of hospitals in North Carolina suggests there may indeed be reason for concern; the bulk of hospital expenditures were for financial assistance to individual patients rather than for activities aimed to improve community health or address social determinants of health (Fos et al. 2019). Another study suggested the irony that hospitals in areas of very high health needs may face demands to devote more resources to meeting patient needs that reduce their ability to contribute to the overall health of their communities (Singh et al. 2018).

5.13.3 Internet Search Engines: Google

When Google launched "Google Flu Trends" in 2008 and claimed to have predicted influenza activity in advance of public health, it caught public attention. The algorithm later required adjustment, however, because of changes in search queries (Cook et al. 2011), and was discontinued. But interest remains in the improved use of internet search engine data with machine learning to detect health trends such as influenza outbreaks. For example, a collaboration between researchers in Mexico and researchers in Boston has recently developed models for predicting influenza in several Latin American countries where disease burdens are high and prediction therefore quite useful (Clemente et al. 2019).

Google itself continues to be interested in health surveillance. DeepMind Health is the medical unit of an AI company acquired by Google in 2014; it was brought directly into Google in 2018. DeepMind has contracts with the British National Health Service, raising concerns among observers that Google will have unprecedented access to patient health data (Stokel-Walker 2018). An earlier agreement between DeepMind and the NHS was determined by the Information Commissioner's Office to have violated patient data protection regulations (ICO 2017a). Under that agreement, Royal Free Hospital in London had agreed to share identifiable data about 1.6 million patients to test the clinical safety of an application, Streams, used to identify risks for acute kidney injury. Royal Free had believed that the use of the data qualified as processing for direct patient care and had not required that the data be anonymized. Royal Free also had not been transparent with patients about the use of the data. The Information Commissioner found that this was a use of personal data that patients would not reasonably expect and that the data use was not what was minimally necessary to achieve the purpose of the safety testing (ICO 2017b). The data use was thus determined to have violated a number of the Caldicott principles discussed earlier in this chapter.

5.13.4 Social Media: The Facebook Example

Many observers now regard the ability to understand what people have in common as a signal achievement of social media. Social media sites such as Facebook have developed algorithms that create pathways to link people together. These algorithms can be adjusted by machine learning techniques and refined in specific areas in considerable detail. But they have of course also brought much criticism of social media. Facebook is at the center of these controversies, as it is the largest social media site with over two and a half billion users in 2020. Facebook also owns Instagram and WhatsApp, among many other companies.

Facebook uses algorithms to prioritize timelines, send news feeds, and serve content from advertisers. In 2018, in response to criticism that they were circulating propaganda or fake news that may have distorted electoral judgments, and that they were prioritizing brand advertising over meaningful interactions with friends, Facebook adopted a new algorithm for structuring news distributions to users. The algorithm itself is protected intellectual property and not open to scrutiny, but according to Montells (2019) appears to prioritize factors such as how frequently a friend has visited another friend's page (affinity), the quality of the post for generating clicks (clickbait), the interactions with the post by others in the network, and the recency of the post.

Through algorithm mediated interactions, Facebook can realize—or undermine—public health goals in several ways. Health information is frequently shared through news feeds. Facebook has announced that it would try to stop misinformation about vaccinations from circulating through its news feeds; it will no longer recommend anti-vaccination groups or allow advertising that contains false information about vaccines (Thebault 2019). It is taking similar steps against groups that post misinformation about supposed cancer cures such as baking soda, apple cider vinegar, or frankincense, and supposed cures for autism such as swallowing bleach (Ohlheiser 2019). However, these steps only downrank misinformation and groups sharing it. Facebook continues to host many such groups and simple google searches will reveal them to interested users (e.g. VRM 2019).

For COVID-19, Facebook has become somewhat more aggressive against misinformation. In March 2020, the site launched an "Information Center" which it features at the top of the news feed to convey accurate information about the pandemic. A month later, it extended fact-checking, attached warnings to debunked stories, reduced the distribution of these stories, removed misinformation that could cause imminent physical harm (such as that drinking bleach could cure COVID-19), and began sending news feed messages to people who had interacted with such harmful misinformation (Rosen 2020). Removing misinformation and sending warnings to people who might have accepted it represents new steps for Facebook.

On Facebook, friends may also share news about their own health or the health of others they know. News feeds will report happy news such as births or a child's first steps. They will also reveal less happy news such as daily reports of a grueling course of chemotherapy or shock at a recent diagnosis. People can also join groups devoted to specific health issues. Some of these groups are open membership while others are "private" groups requiring approval for membership. Groups can also be hidden, so that they can only be found by their members.

All of this information—what people say, who people friend, what groups people join, and what information attracts attention—can be used for surveillance. The information in social media sites such as Facebook may be particularly useful because it may contain not only information about individual health conditions and health trends, but because it will also contain information about attitudes towards health and health care. Through Facebook, public health may not only learn whether someone has influenza, but also whether somebody—or many somebodies—is taking the possibility of an influenza outbreak seriously. Research has also suggested that data from social media sites such as Twitter, in combination with other data digital data streams such as reports from smart thermometers, may facilitate providing early warning of emerging outbreaks of COVID-19 (Kogan et al. 2020).

Facebook's algorithm may also be put to service in making inferences about the health situation of particular individuals. Facebook now engages in suicide prevention efforts that have caught national attention. The company will call emergency services if it believes a user may be at very high risk because of the user's interactions on its platform. It uses an algorithm to assess risk; although it has offered some details on the algorithm, the full algorithm has not been made public. Ian Barnett and John Torous (2019) have recently criticized this limited transparency for medical research, pointing out that the credentials of Facebook's Community Operations team who reviews these data and the outcomes from approximately 3500 notifications to local emergency services to date are not fully clear. Barnett and Torous also ask whether a practice such as this should more properly be for public health:

> Facebook's suicide prevention efforts leads to the question of whether this falls under the scope of public health. Considering the amount of personal medical and mental health information Facebook accumulates in determining whether a person is at risk for suicide, the public health system it actives through calling emergency services, and the need to ensure equal access and efficacy if the system does actually work as hoped, the scope seems more fitting for public health departments than a publicly traded company whose mandate is to return value to shareholders. What happens when Google offers such a service based on search history, Amazon on purchase history, and Microsoft on browsing history? In an era where integrated mental health care is the goal, how do we prevent fragmentation by uncoordinated innovation? (565)

Youth suicide is a major public health problem in the U.S. and Facebook's efforts may be beneficial. But it is also easy to imagine that they might prove quite intrusive. If the predictive algorithms are wrong, especially if they lead to false positives in a low incidence population, the results may be very disturbing. And it is not hard to imagine algorithms such as these being put to many other uses: trying to detect whether someone is at risk of committing violent acts against others, coming down with a dangerous infection, experiencing an opioid overdose, or even experiencing early signs of an undetected cancer.

All of these uses of social media have potential benefits for individuals or for those they might harm. But they are deeply troubling as well. The inequities of predictive algorithms are well known; algorithms can yield skewed judgments if the data themselves are skewed in some way. For example, many have argued that predictive policing or detention algorithms in use in the U.S. are skewed by race and that algorithms predicting the likelihood of addiction may be skewed by poverty (Eubanks 2018). There is little or no transparency about how these algorithms work because they are protected intellectual property.

Finally, in the U.S. especially there is limited oversight of how social media may be developing public health functions. Social media sites are commercial entities with interests that may not always align with the public interest. Their support through advertising in particular may influence how they are structured, how they appeal to users, and how they harvest and use data shared through their sites.

5.14 Summary

Many of these kinds of data—patient medical records, bloodspots retained from newborn screening, and some tissue samples in biobanks, some information in registries, and some information from research—were initially produced in the course of individual health care. Whether individuals are likely to expect information from their health care to be assembled and repurposed for public health is not clear. Individuals may expect their records to be used for their health care and perhaps for the improvement of health care by the providers from which they have received treatment. They are less likely to expect, however, that information gleaned in their care might be used to serve the public more generally, even when the use is in the service of health overall. This is especially true when the information concerns children or others who may be unaware that they even had the medical care that was the source of the information. The calls for destruction of bloodspots retained in newborn screening are a cautionary tale for what may be the result of what people regard as unfair surprise.

We would not necessarily draw the conclusion, however, that in most cases individual consent should be required for public health use of data gained in medical care. Rather, these are uses that must be considered in light of the considerations we raised at the outset of this volume. Openness and transparency are particularly important to avoid surprises such as those with newborn screening. As we have suggested, electronic means of communication available for use with patient records may make it far easier to communicate with patients about public health uses of medical records and their benefits. In addition, public health uses of information from medical care must not increase risks to patients. When public health uses information drawn from medical care, it must be subject to the same standards that are applied to these medical records for data security and disclosures. Public health must not access identifiable patient information without performing adequate security risk assessments and implementing appropriate protections against security threats. Further, public health must not use information in ways that significant numbers of people would find objectionable. For example, the British Wellcome Trust commissioned a study of patient attitudes towards commercial access to health records. The study found that participants were more likely to accept the access when it was for clear public benefit rather than for private benefit and that uses such as for medical research in the public interest could be acceptable while uses for insurance underwriting or for marketing would not be (Ipsos MORI 2016). The study also found that data security was essential to the public. Public health should also attend to equity issues in the use of data it has gained from medical care. There should also be oversight of public health attention to these concerns, as with the Caldecott guardian structure in the UK.

New data, new users, and new analytic methods will continue to present ethical challenges for public health. In weighing the ethics of these uses, public health will need to consider what these uses and users may contribute to the enterprise of improving overall health. Enhanced risks and ways of mitigating them will need to be identified. Public reactions of surprise are likely if data are repurposed for goals they do not accept or if the data collection or use seems unduly intrusive or threatening. Equity is an ongoing challenge. In general, public health will need to do all that it can to ensure that the ethical considerations we put forth in the beginning of this volume are addressed in ways that can enhance trust and thus sustain surveillance.

References

23andMe. 2019. Do You Speak BRCA? https://www.23andme.com/brca/. Accessed 24 July 2020.

Aarestrup, F.M., A. Albeyatti, W.J. Armitage, et al. 2020. Towards a European Health Research and Innovation Cloud (HRIC). *Genome Medicine* 12: 18.

Association of State and Territorial Health Officials (ASTHO). 2012. Maximizing the Community Health Impact of Community Health Needs Assessments Conducted by Tax-exempt Hospitals. http://www.astho.org/Programs/Access/Community-Health-Needs-Assessment/Consensus-Statement/. Accessed 29 July 2020.

Barnett, Ian, and John Torous. 2019. Ethics, Transparency, and Public Health at the Intersection of Innovation and Facebook's Suicide Prevention Efforts. *Annals of Internal Medicine* 170 (8): 565–566.

Benkler, Yochai. 2019. Don't Let Industry Write the Rules for AI. *Nature* 569 (7755): 161.

Blount, Carrie, Cate Walsh Vockley, Amy Gaviglio, Michelle Fox, Brook Croke, Lori Williamson Dean, and The Newborn Screening Task Force on behalf of the NSGC Public Policy Committee. 2014. Newborn Screening: Education, Consent, and the Residual Blood Spot., The Position of the National Society of Genetic Counselors. *Journal of Genetic Counseling* 23 (1): 16–19.

Blue Frog Robotics. 2019. Buddy for Senior Citizen(s) Who Live at Home. http://www.blue-frogrobotics.com/en/buddy-elderly-senior-robot-alone/. Accessed 29 July 2020.

Brief for the Federal Respondents. 2020. *California v. Texas*, Nos. 19-840 and 19-1019. United States Supreme Court. https://www.supremecourt.gov/DocketPDF/19/19-840/146406/20200625205555069_19-840bsUnitedStates.pdf. Accessed 24 July 2020.

Brinkel, Johanna, Alexander Krämer, Ralf Krumkamp, Jürgen May, and Julius Fobil. 2014. Mobile PhoneBased mHealth Approaches for Public Health Surveillance in Sub-Saharan Africa: A Systematic Review. *International Journal of Environmental Research and Public Health* 11 (11): 11559–11582.

Bunnik, Eline M., A. Cecile, J.W. Janssens, and Maarje H.N. Schermer. 2014. Informed Consent in Direct-to-Consumer Personal Genome Testing: The Outline of a Model Between Specific and Generic Consent. *Bioethics* 28 (7): 343–351.

Carlisle, Nate. 2019. Utah's Controlled Substance Database susceptible to hacking, auditors warn. The Salt Lake Tribune (February 13). https://www.sltrib.com/news/politics/2019/02/13/utahs-controlled/. Accessed Feb 20, 2021.

Carpenter v. United States. 2018. https://www.supremecourt.gov/opinions/17pdf/16-402_h315.pdf. Accessed July 29, 2020.

Chang, Hsiao-Han, Amy Wesolowski, Ipsita Sinha, Christopher G. Jacob, Ayesha Mahmud, Didar Uddin, Sazid Ibna Zaman, Md Amir Hossain, M. Abul Faiz, Aniruddha Ghose, Abdullah Abu Sayeed, M. Ridwanur Rahman, Akramul Islam, Mohammad Jahirul Karim, M. Kamar Rezwan, Abul Khair Mohammad Shamsuzzaman, Sanya Tahmina Jhora, M.M. Aktaruzzaman, Eleanor Drury, Sonia Gonçalves, Mihir Kekre, Mehul Dhorda, Ranitha Vongpromek, Olivo Miotto, Kenth Engø-Monsen, Dominic Kwiatkowski, Richard J. Maude, and Caroline Buckee. 2019. Mapping Imported Malaria in Bangladesh Using Parasite Genetic and Human Mobility Data. *Epidemiology and Global Health. eLife* 8: e43481. https://doi.org/10.7554/eLife.43481. Accessed 29 July 2020.

Chattopadhyay, Ishanu, Emre Kiciman, Joshua W. Elliott, Jeffrey L. Shaman, and Andrey Rzhetsky. 2018. Conjunction of Factors Triggering Waves of Seasonal Influenza. *Epidemiology and Global Health. eLife* 7: 330756. https://doi.org/10.7554/eLife.30756. Accessed 29 July 2020.

Chan, Tom, Concetta Tania Di Iorio, Simon de Lusignan, Daniel Lo Russo, Craig Kuziemsky, and Siaw-Teng Liaw. 2016. The UK National Data Guardian for Health Care's Review of Data Security, Consent and Opt-outs: Leadership in Balancing Public Health with Rights to Privacy? *Journal of Innovation in Health Informatics* 23 (3): 627–632.

Clayton, Ellen W., Colin M. Halverson, Nila A. Sathe, and Bradley A. Malin. 2018. A Systematic Literature Review of Individuals' Perspectives on Privacy and Genetic Information in the United States. *PLoS One* 13 (10): e0204417.

Clemente, Leonardo, Fred Lu, and Mauricio Santillana. 2019. Improved Real-Time Influenza Surveillance: Using Internet Search Data in Eight Latin American Countries. *JMIR Public Health Surveillance* 5 (2): e12214.

Cook, Samantha, Corrie Conrad, Ashley L. Fowlkes, and Matthew H. Mohebbi. 2011. Assessing Google Flu Trends Performance in the United States During the 2009 Influenza Virus A (H1N1) Pandemic. *PLOS One*. https://doi.org/10.1371/journal.pone.0023610. Accessed 29 July 2020.

Coyle, D. Tyler. 2017. Sentinel System Overview. https://www.fda.gov/media/110795/download. Accessed 29 July 2020.

Curfman, Gregory D. 2017. All-Payer Claims Databases After *Gobille*. *Health Affairs Blog* (Mar 3). https://www.healthaffairs.org/do/10.1377/hblog20170303.058995/full/. Accessed 22 July 2020.

digitalhealth. 2016. Caldicott 3: Easy to Say, Hard to Do? (Oct 6). https://www.digitalhealth.net/2016/10/caldicott-3-easy-to-say-hard-to-do/. Accessed 23 July 2020.

Dillard v. O'Kelley. 961 F.3d 1048 (8th Cir. 2020).

Editorial. 2011. There Will Be Blood. *Nature* 475 (7355): 139.

Etchegary, Holly, Stuart G. Nicholls, Laure Tessier, Charlene Simmonds, Beth K. Potter, Jamie C. Brehaut, Daryl Pullman, Robyn Hayeems, Sari Zelenietz, Monica Lamoureaux, Jennifer Milburn, Lesley Turney, Pranesh Chakraborty, and Brenda Wilson. 2016. Consent for Newborn Screening: Parents' and Healthcare Professionals' Experiences of Consent in Practice. *European Journal of Human Genetics* 24 (11): 1530–1534.

Eubanks, Virginia. 2018. *Automating Inequality: How High-Tech Tools Profile, Police, and Punish the Poor*. New York: St. Martin's Press.

European Union (EU). 2019. Recital 54: Processing of Sensitive Data in Public Health Sector. https://gdpr-info.eu/recitals/no-54/. Accessed 29 July 2020.

Fairchild, Amy L., Daniel Wolfe, James Keith Colgrove, and Ronald Bayer. 2007. *Searching Eyes: Privacy, the State, and Disease Surveillance in America*. Berkeley: University of California Press.

Ferguson, Andrew Guthrie. 2017. *The Rise of Big Data Policing: Surveillance, Race, and the Future of Law Enforcement*. New York: NYU Press.

Floridi, Luciano. 2014. Open Data, Data Protection, and Group Privacy. *Philosophy & Technology* 27: 1–3.

Food and Drug Administration (FDA). 2017. FDA Approves Pill with Sensor That Digitally Tracks If Patients Have Ingested Their Medication. https://www.fda.gov/news-events/press-announcements/fda-approves-pill-sensor-digitally-tracks-if-patients-have-ingested-their-medication. Accessed July 29, 2020.

Fos, Elmer, Michael Thompson, Christine Elnitsky, and Elena Platonova. 2019. Community Benefit Spending Among North Carolina's Tax-Exempt Hospitals After Performing Community Health Needs Assessments. *Journal of Public Health Management and Practice* 25 (4): E1–E8.

Francis, Leslie P. 2014. Adult Consent to Continued Participation in Patient Registries. *St Louis University Journal of Health Law & Policy* 7 (2): 389–405.

Francis, John G., and Leslie P. Francis. 2014. Privacy, Confidentiality, and Justice: Using Large-Scale Sets of Health Data to Improve Public Health. *Journal of Social Philosophy* 45 (3): 408–431.

Francis, Leslie P., and Michael Squires. 2018. Patient Registries and Their Governance: A Pilot Study and Recommendations. *Indiana Health Law Review* 16: 43–65.

Francis, Leslie P., Margaret P. Battin, Jay Jacobson, and Charles B. Smith. 2009. Syndromic Surveillance and Patients as Victims and Vectors. *Journal of Bioethical Inquiry* 6 (2): 187–195.

Friedler, Sorelle A., Carlos Scheidegger, and Suresh Ventatasubramanian. 2016. On the (im)possibility of Fairness. https://arxiv.org/pdf/1609.07236.pdf?source=post_page%2D%2D%2D% 2D%2D%2D%2D%2D%2D%2D%2D%2D%2D%2D%2D%2D%2D%2D%2D%2D%2D%2D%2 D%2D%2D%2D%2D-. Accessed 27 July 2020.

Friedman, Daniel J., R. Gibson Parrish, and David A. Ross. 2013. Electronic Health Records and US Public Health: Current Realities and Future Promise. *American Journal of Public Health* 103 (9): 1560–1567.

General Data Protection Regulation (GDPR). 2020. https://gdpr-info.eu/. Accessed 21 July 2020.

Gilmartin, Chris, Edward H. Arbe-Barnes, Michael Diamond, Sasha Fretwell, Euan McGivern, Myrto Vlazaki, and Liment Zhu. 2018. Varsity Medical Ethics Debate 2018: Constant Health Monitoring—The Advance of Technology into Healthcare. *Philosophy, Ethics, and Humanities in Medicine* 13: 12. https://doi.org/10.1186/s13010-018-0065-0. Accessed 29 July 2020.

Gobeille v. Liberty Mutual Insurance Company, 577 U.S. ___, 136 S. Ct. 936 (2016).

Google. 2020. Privacy Policy (updated July 1). https://policies.google.com/privacy?hl=en-US. Accessed 29 July 2020.

Guerrini, Christi J., Jill O. Robinson, Devan Petersen, and Amy L. McGuire. 2018. Should Police Have Access to Genetic Genealogy Databases? Capturing the Golden State Killer and Other Criminals Using a Controversial New Forensic Technique. *PLOS Biology* (Oct 2). https://journals.plos.org/plosbiology/article?id=10.1371/journal.pbio.2006906. Accessed 29 July 2020.

Gullo, Karen. 2016. Surveillance Chills Speech—As New Studies Show—And Free Association Suffers. https://www.eff.org/deeplinks/2016/05/when-surveillance-chills-speech-new-studies-show-our-rights-free-association. Accessed 29 July 2020.

Hart, Alexa, Michael Petros, Joel Charrow, Claudia Nash, and Catherine Wicklund. 2015. Storage and Use of Newborn Screening Blood Specimens for Research: Assessing Public Opinion in Illinois. *Journal of Genetic Counseling* 24 (3): 482–490.

Hayeems, Robin Z., Fiona A. Miller, Carolyn J. Barg, Yvonne Bombard, Celine Cressman, Michael Painter-Main, Brenda Wilson, Julian Little, Judith Allanson, Denise Avard, Yves Giguere, Pranesh Chakraborty, and June C. Carroll. 2016. Using Newborn Screening Bloodspots for Research: Public Preferences for Policy Options. *Pediatrics* 137 (6): e20154143.

Health Insurance Portability and Accountability Act Rules (HIPAA). 2020. 45 C.F.R. Parts 160, 162, and 164. https://www.hhs.gov/sites/default/files/ocr/privacy/hipaa/administrative/combined/hipaa-simplification-201303.pdf. Accessed 23 July 2020.

Hoffman, Sharona. 2018. Big Data Analytics: What Can Go Wrong. *Indiana Health Law Review* 15: 227–246.

Hoffman, Sharona, and Andy Podgurski. 2013. Big Bad Data: Law, Public Health, and Biomedical Databases. *Journal of Law, Medicine and Ethics* 41 (Suppl. 1): 56–60.

Information Commissioner's Office (ICO). 2017a. Royal Free – Google DeeMind Trial Failed to Comply with Data Protection Law (July 3). https://ico.org.uk/about-the-ico/news-and-events/news-and-blogs/2017/07/royal-free-google-deepmind-trial-failed-to-comply-with-data-protection-law/. Accessed 29 July 2020.

———. 2017b. Letter from Elizabeth Denham, ICO Commissioner, to Sir David Sloman (July 3). https://ico.org.uk/media/2014353/undertaking-cover-letter-revised-04072017-to-first-person.pdf. Accessed 29 July 2020.

Institute of Medicine (IOM). 2010. *Challenges and Opportunities in Using Residual Newborn Screening Samples for Translational Research*. Washington, DC: National Academies Press.

Intermountain Healthcare. 2019. Intermountain Biorepository. https://intermountainhealthcare.org/about/transforming-healthcare/innovation/intermountain-biorepository/about/. Accessed 29 July 2020.

Internal Revenue Service (IRS). 2018. Community Health Needs Assessment for Charitable Hospital Organizations—Section 501(r)(3) (Nov 18). https://www.irs.gov/charities-non-profits/community-health-needs-assessment-for-charitable-hospital-organizations-section-501r3. Accessed 29 July 2020.

Ipsos MORI. 2016. The One-Way Mirror: Public Attitudes to Commercial Access to Health Data. https://www.ipsos.com/sites/default/files/publication/5200-03/sri-wellcome-trust-commercial-access-to-health-data.pdf. Accessed 29 July 2020.

Kaiser, Jocelyn. 2018. We Will Find You: DNA Search Used to Nab Golden State Killer Can Home in on About 60% of White Americans. *Science* (Oct 11) [online]. https://www.sciencemag.org/news/2018/10/we-will-find-you-dna-search-used-nab-golden-state-killer-can-home-about-60-white. Accessed 29 July 2020.

Kogan, Nicole E., Leonardo Clemente, Parker Liautaud, Justin Kaashoek, Nicholas B. Link, Andre T. Nguyen, Fred S. Lu, Peter Huybers, Bernd Resch, Clemens Havas, Andreas Petutschnig, Jessica Davis, Matteo Chinazzi, Backtosch Mustafa, William P. Hanage, Alessandro Vespignani, and Mauricio Santillana. 2020. An Early Warning Approach to Monitor COVID-19 Activity with Multiple Digital Traced in Near Real-Time. (July 7). https://arxiv.org/pdf/2007.00756.pdf. Accessed 24 July 2020.

Kostkova, Patty. 2018. Disease Surveillance Data Sharing for Public Health: The Next Ethical Frontiers. *Life Sciences, Society and Policy* 14: 16–21.

Kruse, Clemens Scott, Anna Stein, Heather Thomas, and Harmander Kaur. 2018. The us of Electronic Health Records to Support Population Health: A Systematic Review of the Literature. *Journal of Medical Systems* 42 (11): 214.

Kupferschmidt, Kai. 2019. The Planet's Premier Health Agency Has Announced Drastic Reforms. Critics Say They Aren't Drastic Enough. *Science* (Mar 12) [online]. https://www.sciencemag.org/news/2019/03/planet-s-premier-health-agency-has-announced-drastic-reforms-critics-say-they-arent. Accessed 29 July 2020.

Lee, Elizabeth C., Jason M. Asher, Sandra Goldlust, John D. Kraemer, Andrew B. Lawson, and Shweta Bansal. 2016. Mind the Scales: Harnassing Spatial Big Data for Infectious Disease Surveillance and Inference. *The Journal of Infectious Diseases* 214 (Suppl. 4): S409–S413.

Lewis, Michelle Huckaby, and Aaron J. Goldenberg. 2015. Return of Results from Research Using Newborn Screening Dried Blood Samples. *The Journal of Law, Medicine, & Ethics* 43 (3): 559–568.

Lewis, Michelle Huckaby, Michael E. Scheurer, Robert C. Green, and Amy L. McGuire. 2012. Research Results: Preserving Newborn Blood Samples. *Science Translational Medicine* 4 (159): 159cm12.

Ludman, Evette J., Stephanie M. Fullerton, Leslie Spangler, Susan Brown Trinidad, Monica M. Fujii, Gail P. Jarvik, Eric B. Larson, and Wylie Burke. 2010. Glad You Asked: Participants' Opinions of Re-Consent for dbGaP Data Submission. *Journal of Empirical Research on Human Research Ethics* 5 (3): 9–16.

Mariner, Wendy K. 2007. Mission Creep: Public Health Surveillance and Medical Privacy. *Boston University Law Review* 87: 347–396.

McLaughlin, Robert H., Christina A. Clarke, LaVera M. Crawley, and Sally L. Glaser. 2010. Are cancer registries unconstitutional? *Social Science & Medicine* 70 (9): 1295–1300.

McLennan, Stuart, Leo Anthony Celi, and Alena Buyx. 2020. COVID-19: Putting the General Data Protection Regulation to the Test. *JMIR Public Health Surveillance* 6 (2): e19279.

Menderski, Maggie. 2014. Superfund Legacy Remains in Taylorville and Beyond. *The State Journal Register* (Sept 13) [online], https://www.sj-r.com/article/20140913/NEWS/140919731. Accessed 29 July 2020.

Meyers, Jessica M. 2019. Artificial Intelligence and Trade Secrets. American Bar Association. *Landslide* II (3). https://www.americanbar.org/groups/intellectual_property_law/publications/landslide/2018-19/january-february/artificial-intelligence-trade-secrets-webinar/. Accessed 27 July 2020.

Michigan Department of Health & Human Services (Michigan DHHS). 2019. Michigan BioTrust for Health—Consent Options. https://www.michigan.gov/mdhhs/0,5885,7-339-73971_4911_4916_53246-244016%2D%2D,00.html. Accessed 29 July 2020.

Mittelstadt, Brent Daniel, and Luciano Floridi. 2016. The Ethics of Big Data: Current and Foreseeable Issues in Biomedical Contexts. *Science and Engineering Ethics* 22: 303–341.

Molnár-Gábor, Fruzsina. 2018. Germany: A Fair Balance Between Scientific Freedom and Stat Subjects' Rights? *Human Genetics* 137 (8): 619–626.

Montells, Laura. 2019. How Facebook Algorithm Works. *Metricool* [online]. https://metricool.com/what-is-facebook-edgerank-or-how-facebook-algorithm-works/. Accessed 24 July 2020.

National Cancer Institute (NCI). 2021. *What is a Cancer Registry?* https://seer.cancer.gov/registries/cancer_registry/index.html

National Center for Biotechnology Information (NCBI). 2019. dbGap. https://www.ncbi.nlm.nih.gov/gap/. Accessed 29 July 2020.

National Data Guardian for Health and Care (NDG). 2016. *Review of Data Security, Consent and Opt-Outs* (Caldicott 3). https://assets.publishing.service.gov.uk/government/uploads/system/uploads/attachment_data/file/535024/data-security-review.PDF. Accessed 29 July 2020.

Nordfalk, Francisca, and Claus Thorn Ekstrøm. 2019. Newborn Dried Blood Spot Samples in Denmark: The Hidden Figures of Secondary Use and Research Participation. *European Journal of Human Genetics* 27 (2): 203–210.

Office for the National Data Guardian & Connected Health Cities. 2018. *Reasonable Expectations Report.* https://www.connectedhealthcities.org/wp-content/uploads/2018/08/Reasonable-expectations-jury-report-v1.0-FINAL-09.08.18.pdf. Accessed 29 July 2020.

Ohlheiser, Abby. 2019. Facebook Wants to Limit the Reach of Bogus Medical 'Cures' by Treating Them like Spam. *The Washington Post* (July 2) [online]. https://www.washingtonpost.com/technology/2019/07/02/facebook-wants-limit-reach-bogus-medical-cures-by-treating-them-like-spam/?utm_term=.106ea94a95a5. Accessed 29 July 2020.

Orphan Disease Center (ODC). 2020. Patient Registries. https://orphandiseasecenter.med.upenn.edu/patient-registries. Accessed 29 July 2020.

Pasquale, Frank. 2016. *The Black Box Society: The Secret Algorithms That Control Money and Information.* Cambridge, MA: Harvard University Press.

patientslikeme. 2020. Informed Consent to Take Part in patientslikeme digialme™ ignite. https://www.patientslikeme.com/digitalme/consent_document. Accessed 28 July 2020.

Peak, Corey M., Amy Wesolowski, Elisabeth zu Erbach-Schoenberg, Andrew J. Tatem, Erik Wetter, Xin Lu, Daniel Power, Elaine Weidman-Grunewald, Sergio Ramos, Simon Moritz, Caroline O. Buckee, and Linus Bengtsson. 2018. Population Mobility Reductions Associated with Travel Restrictions During the Ebola Epidemic in Sierra Leone: Use of Mobile Phone Data. *International Journal of Epidemiology* 47 (5): 1562–1570.

Platt, Jodyn E., Tevah Platt, Daniel Thiel, and Sharon L.R. Kardia. 2013. 'Born in Michigan? You're in the Biobank': Engaging Population Biobank Participants Through Facebook Advertisements. *Public Health Genomics* 16: 145–158.

Ploug, Thomas, and Søren Holm. 2016. Meta Consent – A Flexible Solution to the Problem of Secondary Use of Health Data. *Bioethics* 30 (9): 721–732.

Rosen, Guy. 2020. An Update on Our Work to Keep People Informed and Limit Misinformation About COVID-19. Facebook (April 20, updated May 12). https://about.fb.com/news/2020/04/covid-19-misinfo-update/. Accessed 24 July 2020.

Saks, Michael J., Adela Grando, Chase Millea, and Anita Murcko. 2020. Advancing the Use of HIE Data for Research. *Arizona State Law Journal* 52: 145–189.

Schechter, Michael S., Aliza K. Fink, Karen Homa, and Christopher H. Goss. 2014. The Cystic Fibrosis Foundation Patient Registry as a Tool for Use in Quality Improvement. *British Medical Journal Quality and Safety* 23: i9–i14.

Shabani, Mahsa, and Pascal Borry. 2018. Rules for Processing Genetic Data for Research Purposes in View of the New EU General Data Protection Regulation. *European Journal of Human Genetics* 26 (2): 149–156.

Singh, Simone R., Geri R. Cramer, and Gary J. Young. 2018. The Magnitude of a Community's Health Needs and Nonprofit Hospitals' Progress in Meeting Those Needs: Are We Faced With a Paradox? *Public Health Reports* 133 (1): 75–84.

Sørensen, Eigil. 2018. Challenges for the World Health Organization. *Tidsskr Nor Legeforen* 138 (1). https://doi.org/10.4045/tidsskr.17.0412. Accessed 5 Aug 2020.

Southern Illinoisan v. Illinois Department of Public Health, 844 N.E.2d 1 (Ill. 2006).

Statens Serum Institut. 2020. The Danish Biobank Register. https://www.danishnationalbiobank.com/danish-biobank-register. Accessed 28 July 2020.

Staunton, Ciara, Santa Slokenberga, and Deborah Mascalzoni. 2019. The GDPR and the Research Exemption: Considerations on the Necessary Safeguards for Research Biobanks. *European Journal of Human Genetics* 27: 1159–1167.

Sterling, Robyn L. 2011. Genetic Research Among the Havasupai: A Cautionary Tale. *AMA Journal of Ethics (Virtual Mentor)* 13 (2): 113–117.

Stokel-Walker, Chris. 2018. Why Google Consuming DeepMind Health Is Scaring Privacy Experts. *Wired* (Nov 18) [online]. https://www.wired.co.uk/article/google-deepmind-nhs-health-data. Accessed 29 July 2020.

Sweeney, Latanya, Michael von Loewenfeldt, and Melissa Perry. 2018. Saying It's Anonymous Doesn't Make It So: Re-identifications of "Anonymized" Law School Data. *JOTS Technology Science* (Nov 13) [online]. https://techscience.org/a/2018111301/. Accessed 27 July 2020.

Taylor, Kyle, and Laura Silver. 2019. Smartphone Ownership Is Growing Rapidly Around the World, But Not Always Equally. Pew Research Center (Feb 5). https://www.pewresearch.org/global/wp-content/uploads/sites/2/2019/02/Pew-Research-Center_Global-Technology-Use-2018_2019-02-05.pdf. Accessed 20 Feb 2021.

Thebault, Reis. 2019. Facebook Says It Will Take Action Against Anti-vaccine Content. Here's How It Plans to Do It. *The Washington Post* (Mar 7) [online]. https://www.washingtonpost.com/business/2019/03/07/facebook-says-it-will-take-action-against-anti-vax-content-heres-how-they-plan-do-it/?noredirect=on&utm_term=.063f6b2ead69. Accessed 29 July 2020.

Thiel, Daniel B., Tevah Platt, Jodyn Platt, Susan B. King, and Sharon L.R. Kardia. 2014. Community Perspectives on Public Health Biobanking: An Analysis of Community Meetings on the Michigan BioTrust for Health. *Journal of Community Genetics* 5 (2): 125–138.

UK Biobank. 2018. Information About the New General Data Protection Regulation (GDPR). https://www.ukbiobank.ac.uk/wp-content/uploads/2018/10/GDPR.pdf. Accessed 24 July 2020.

Understanding Patient Data. 2018. Why an Opt-out Rather than an Opt-in or Consent? (May 25). https://understandingpatientdata.org.uk/news/why-an-opt-out. Accessed 29 July 2020.

Vaccine Resistance Movement (VRM). 2019. Updates and News. https://www.facebook.com/groups/VaccineResistanceMovement. Accessed 29 July 2020.

Vandemeulebroucke, Tijs, Bernadette Dierckx de Casterlé, and Chris Gastmans. 2018. The Use of Care Robots in Aged Care: A Systematic Review of Argument-Based Ethics Literature. *Archives of Gerontology and Geriatrics* 74: 15–25.

Wesolowski, Amy, Nathan Eagle, Andrew J. Tatem, David L. Smith, Abdisalan M. Noor, Robert W. Snow, and Caroline O. Buckee. 2012. Quantifying the Impact of Human Mobility on Malaria. *Science* 338 (6104): 276–270.

Whitehurst, Lindsay. 2017. APNewsBreak: Utah to obey order for DEA drug database search (August 22). https://apnews.com/article/a0985c75b8ce43c8a834131a42b7fdbc. Accessed Feb 20, 2021.

Whalen v. Roe, 429 U.S. 587 (1977).

Williams, Karmen S., and Gulzar H. Shah. 2016. Electronic Health Records and Meaningful Use in Local Health Departments: Updates From the 2015 NACCHO Informatics Assessment Survey. *Journal of Public Health Management and Practice* 22 (Suppl. 6): S27–S33.

World Health Organization (WHO). 2016. Framework of Engagement with Non-State Actors (May 28). http://apps.who.int/gb/ebwha/pdf_files/wha69/a69_r10-en.pdf. Accessed 29 July 2020.

————. 2019. Use of Smartphone Technology for Elimination of Malaria in Bhutan. http://www.searo.who.int/entity/health_situation_trends/country_profiles/e_health/mal_bhu/en/. Accessed 29 July 2020.

————. World Health Regulations (2005), 3rd ed. https://apps.who.int/iris/bitstream/handle/10665/246107/9789241580496-eng.pdf?sequence=1. Accessed 29 July 2020.

World Health Organization (WHO) director-general. 2019. Engagement with Non-State Actors: Report by the Director General (Nov 23). http://apps.who.int/gb/ebwha/pdf_files/EB144/B144_36-en.pdf. Accessed 29 July 2020.

Xu, Heng, and Nan Zhang. 2020. Implications of Data Anonymization on the Statistical Evidence of Disparity. Working Paper presented at Privacy Law Scholars Conference (PLSC 2020). SSRN. https://papers.ssrn.com/sol3/papers.cfm?abstract_id=3662612. Accessed 28 July 2020.

Yoo, Ji Su, Alexandra Thaler, Latanya Sweeney, and Jinyan Zang. 2018. Risks to Patient Privacy: A Re-identification of Patients in Maine and Vermont Statewide Hospital Data. *JOTS Technology Science* (Oct 9) [online]. https://techscience.org/a/2018100901/. Accessed 27 July 2020.

Zou, James, and Londa Schiebinger. 2018. AI Can Be Sexist and Racist – It's Time to Make It Fair. *Nature* 559 (7714): 324–326.

Chapter 6
Surveillance for the "New" Public Health

Up to this point in the volume, we have considered these traditional forms of public health surveillance: detecting outbreaks of dangerous contagious disease, case finding and contact tracing, and identification of environmental hazards. We have also discussed the introduction of new data, new actors, and new technologies into public health surveillance. We have located these discussions primarily in the realm of public health missions of containing infectious disease and reducing environmental exposures. But public health now includes far more about the well-being of the population.

6.1 Public Health and Population Well-being

In the last half of the twentieth century, public health increasingly saw itself as having the broader mission of maximizing human health and well-being, a mission aligned with the World Health Organization's definition of health as "a state of complete physical, mental and social well-being and not merely the absence of disease or infirmity" (WHO 2019). Perhaps not accidentally, the time between the end of the Second World War and the advent of the HIV/AIDS epidemic around 1980 was an era of great optimism about the conquest of infectious disease through antimicrobial therapy (Wiley 2012, p. 220). It was also an era of increasing recognition of the role played by social factors in influencing health—that is, the social determinants of health. In the U.S., optimism about the ability of government to promote overall welfare was at its height during the administration of President Lyndon Johnson, who initiated pursuit of the "Great Society" in 1964.

To further the mission of improving population health broadly conceived, public health directed attention to addressing rising rates of chronic or noncommunicable diseases (NCDs) such as type 2 diabetes, heart disease, arthritis, chronic obstructive lung disease, and many cancers. Policy priorities were increased governmental

© Springer Nature Switzerland AG 2021   159
J. G. Francis, L. P. Francis, *Sustaining Surveillance: The Importance of Information for Public Health*, Public Health Ethics Analysis 6,
https://doi.org/10.1007/978-3-030-63928-0_6

support for social safety nets, disease prevention, and primary care. The Declaration of Alma-Ata issued by the 1978 International Conference on Primary Health Care invoked the WHO definition of health to urge governments to formulate policies and plans of action to sustain primary health care as part of a national health care system (WHO 2019a). Importantly, the Declaration asserted the responsibility of government for the general health and well-being of society (Tulchinsky and Varivikova 2010, 2001) These governmental efforts were seen as interconnected with efforts to address individual behaviors judged significantly deleterious to health, such as smoking, poor diet, alcohol consumption, and lack of exercise. Improvement of population health overall thus was judged to require combined efforts by governments and individuals.

This recasting of the mission of public health to improving overall population well-being was dubbed "the new public health." The WHO's Ninth General Programme of Work (1996–2001) recognized the new public health as mobilizing communities, public health, and political leaders around concerted efforts to improve health and well-being overall (Ncayiyana 1995). The Programme also realized, however, that these efforts were encountering resistance to both the language and the scope of the new public health. Resistance took political form, as many countries drew back from supporting or funding increased roles for their governments. Perhaps as a rhetorical strategy, or perhaps as a matter of fundamental changes in the understanding of the etiology of poor health, the new public health did not leave images of contagion behind. Obesity or diabetes were described as "epidemics" for which alarms were to be sounded (WHO 2003). Today, opioid use and the deaths that have resulted are officially labeled an "epidemic" and a "public health emergency" by the US government (DHHS 2019).

The extent to which the new public health has become the norm is also illustrated by the United Nations sustainable development goals for 2030, which lists "good health and well-being" as Goal 3. Among Goal 3 targets are not only ending the epidemics of AIDS and tropical diseases and reducing deaths and ill-health from hazardous pollution, but also strengthening prevention and treatment of substance abuse and halving the number of deaths and injuries from road traffic accidents. The target of supporting development of vaccines and medicines for conditions primarily affecting developing countries includes both communicable and non-communicable diseases alike (UN 2015). The WHO Global Action Plan in support of these sustainable development goals accelerates efforts to address the social determinants of health (WHO 2019b). While the WHO does not have enforcement authority in the sense that it can act as a health police for the new public health, it does establish priorities for observation and funding that are in turn meant to be influential on the priorities of states parties. In the United States, Healthy People 2020 Leading Health Indicators include nutrition, physical activity, and obesity; substance abuse; and tobacco use (ODPHP 2020). Healthy People 2020 also devotes a separate topic to the social determinants of health, with a leading indicator of students entering 9th grade graduating from high school within 4 years. Assessing progress on new public health efforts requires surveillance of a wide range of factors from weight to education, far beyond the communicable diseases and environmental subjects of earlier public health.

6.2 Surveillance for the New Public Health

Primary surveillance goals of the new public health include the incidence and prevalence of NCDs such as diabetes; the frequency and distribution of contributing behaviors such as smoking, diet, and exercise; and the relative distribution of social factors such as education, housing, employment, or economic inequality. Put most generally, the new public health widens the scope of surveillance beyond the incidence and prevalence of contagious and toxic disease to a broad range of social factors that may affect health and well-being. For obesity, these factors might reach beyond weight to include food deserts, safe playgrounds, and cultural norms. For opioids, they might include chronic pain, disability, limited education, unemployment, and other factors implicated in rising rates of "deaths of despair" (Case and Deaton 2015).

At the international level, WHO structures its data collection in light of the UN sustainable development goals. It lists seventeen different types of health and health-related target indicators, including as major categories NCDs, substance abuse, road traffic injuries, and tobacco control (WHO 2019c). WHO (2019d) clusters major non-communicable diseases into four broad areas: cardiovascular disease, cancers, diabetes, and chronic respiratory disease. According to WHO, in 2016 these four areas accounted for 71% of all deaths worldwide. In low- and middle-income countries, 48% of deaths from these conditions are premature in the sense that they occur before age 70. WHO contends that these conditions are linked to modifiable risk factors such as tobacco, harmful use of alcohol, unhealthy diet, insufficient physical activity, overweight/obesity, high blood pressure, raised blood sugar, and high cholesterol. It estimates that 80% of these risk factors are modifiable through public policy changes. To take examples of how these risks might be modifiable, fewer than half (46%) of WHO member states have policies controlling alcohol use by age and licensing requirements, only 10% of people live in countries with tax rates on cigarettes that WHO regards as sufficient, and 39% of both men and women in the world are obese or overweight. Dementia is another type of NCD drawing attention from WHO; WHO launched a global monitoring system in 2017 to track progress on national policy, risk reduction, and dementia care and treatment (WHO 2019e).

In the US, surveillance for NCDs, related behavioral factors, and the social determinants of health is a complex mix of efforts by the federal government, state governments, and other sources. The US CDC is the primary agency responsible for surveillance at the federal level; it relies largely on information supplied by agreement with the states and on surveys that it conducts, including the Behavioral Risk Factor Surveillance System (BRFSS), the Youth Risk Behavior Surveillance System (YRBS), the National Health Interview Study (NHIS), and the National Health and Nutrition Examination Survey (NHANES). In comparison to the WHO figures, infectious diseases account for a very small percentage of U.S. mortality, only just over 5% according to the latest available data (Hansen et al. 2016), although this may change with COVID-19. Influenza and pneumonia accounted for over 75% of these deaths from infection. Death rates from HIV/AIDS rose and then fell over the period between 1980 and 2014, largely due to the availability of more successful anti-retroviral treatment. Mortality rates varied significantly by county, however, with respiratory disease death rates higher in the northeast and HIV/AIDS rates higher in the southeast (el Bcheraoui et al. 2018). CDC groups its data statistics by topics; groupings include alcohol use, cancer, diabetes, heart disease, overweight and obesity, physical activity, and tobacco use. NHIS, BRFSS, YRBS, NHANES, and cancer registries are the primary sources for these data.

It is fair to say that political factors also play into decisions about what to surveille. In the United States, while automobile accident death rates are widely known, death rates from guns, although very high, are more difficult to tease out from the available statistics. The CDC maintains a data category for violent deaths which indicated that in 2015 approximately 62,000 persons died by violence. These data are obtained from death certificates, medical examination, law enforcement reports, and secondary sources; 27 states collected statewide data for 2015 but the remainder did not (Jack et al. 2018). Thus, the data are incomplete, although nonetheless telling as far as they go. According to one recent analysis, injury was the leading cause of death for children; 20% of injuries occurred in motor vehicle accidents and 15% were firearm related. Both of these rates were far higher than the rates in other high-income countries; the U.S. auto accident death rate for children was three times higher than the average in comparable countries and the firearm death rate was an astounding 36.5 times higher (Cunningham et al. 2018). Rates in rural areas were higher than in suburban or urban areas, particularly for automobile accidents. Rates of firearm related deaths and accidental deaths overall also were higher for males and for blacks. The authors of this study conclude that application of rigorous public health methods has had a major difference in reducing childhood deaths from automobile accidents and should be applied to other categories of injury-related deaths. An editorial accompanying the study judges that deaths from gun trauma are an "underrecognized public health problem" that has been mischaracterized as deaths due to "accidents" in a way that conveys a "sense of helpless inevitability" (Campion 2018). These statistics are another illustration of the point made in Chapter 4, that ethical issues attend what is not surveilled as much as what is surveilled.

State disease reporting requirements focus primarily on contagious diseases or bioterrorism. To take one example of state notifiable conditions, the California Department of Public Health (2018) lists 80 communicable diseases that includes Ebola, Flu, HIV/AIDS, sexually transmitted disease, Hepatitis B, Zika, tuberculosis, and many other less well-known conditions. These are all conditions that have been judged to have the potential to threaten health if they are not identified in order to interrupt transmission. Reportable NCDs and conditions include disorders characterized by lapses of consciousness, suspected pesticide injuries, and brain tumors. The first of these—lapse of conscience—is important for drivers who might be hazardous to others. The latter two are related to environmental hazards. NCDs such as diabetes or substance use disorder are not among the listed conditions requiring reports. National statistics for NCDs such as these typically rely on the federal surveys conducted by CDC, such as BRFSS, NHANES, and NHIS (e.g. CDC 2020). Fragile funding and response rates for these federal surveys are a continuing problem, however. States also maintain registries for birth defects and cancers in response to funding from the federal government, as described in Chapter 5.

6.3   Libertarianism and Challenges to Surveillance for the New Public Health

The new public health seeks to improve population health and prevent illness of whatever kind (Nuffield Council 2007, v). Its primary justification is consequentialism applied at the health policy level: policies should be designed to promote the good of health. To the extent that differences in social determinants of health such as education, employment, or housing affect overall population health, issues of equity and justice also require attention. These considerations apply at the population level: overall well-being and its distribution throughout the population are the critical aims of the new public health (Parmet 2009).

Not surprisingly, this population-level approach does not sit well with individualist and libertarian political ideologies, especially in the United States (e.g. Wiley 2016; Wiley et al. 2015; Jacobson 2014). In the US, themes emphasized in health policy include individuals' responsibility for their health, development of increasingly individualized forms of health care treatment such as precision medicine, and improved access of individuals to care. The Affordable Care Act supports incentives for wellness programs and smoking cessation that are aimed to encourage individual health improvements rather than addressing social factors influencing health such as job insecurity.

Opposition to the new public health in the U.S. began to flourish during the administration of President Ronald Reagan in the 1980s. The politics in the U.S. had shifted away from confidence in government. Ronald Reagan was elected president in 1980, and with his presidency came sustained efforts to curtail the role of government. Budget cuts enacted in the first year of the Reagan administration vastly reduced benefits for the working poor and reduced the reach of the Medicaid program to provide health insurance for those without resources. Through waiver programs, states also received more flexibility in designing the structure of their Medicaid programs, enabling them to decide whether to cover home and community-based services for people with disabilities or to require recipients to receive certain benefits through managed care plans. Social safety nets frayed with these attacks on what was judged to be welfare-created dependency. Efforts were made to introduce market incentives through changes in reimbursement structures in the Medicare program that serves the elderly and people with disabilities (Ethridge 1983). Another example of the introduction of private sector incentives in the U.S. during the period was the Bayh-Dole Act of 1980, which permits universities and non-profits receiving federal funding for research to maintain ownership of intellectual property rights in discoveries resulting from the research. Support for spending on medical research through the National Institutes of Health continued, however.

A particularly vehement and influential early critic of the new public health was the libertarian law professor Richard Epstein, a member of the faculty at the University of Chicago and well-known for his use of free-market economic tools in analyzing law. A conference at the University of Chicago in 2002 addressed the social determinants of health and resulted in a special issue of the journal *Perspectives in Biology & Medicine* that included Epstein's critique of the new public health and responses to it. In a speech at a conference on obesity held by the conservative-leaning American Enterprise Institute in 2003, Epstein sparked intense controversy when he argued for limiting the role of public health to contagious diseases and what the law calls environmental nuisances—conditions on the property of one person that spill over to harm others. Public health activities aimed at these concerns, Epstein thought, addressed situations in which governmental intervention was needed to avert the market's failure to yield desired results. Because cleaning up the environment involves the production of public goods, as described in Chapter 4, market incentives aimed at individuals cannot be relied on to do the job. Nor can individual actions, coupled with damage remedies when some unreasonably inflict harm on others, succeed in reducing the spread of infection. On Epstein's libertarian view, the government should refrain from interfering with the market, taking action only when the market fails, as it will with infection and pollution. New public health actions designed to discourage unhealthy behaviors such as soda taxes aimed to reduce soft drink consumption fall outside of the proper scope of government, on this approach.

Epstein's justification for his anti-paternalist positions about the limited role of government (2003) invoked consequentialist reasoning directly at odds with arguments offered in favor of the new public health. His argument rested on empirical assumptions of classical liberalism about market functioning and the success of individuals in directing their own lives. Epstein wrote:

> My broad thesis is that the "old" public health is superior to the new, whose broad (and meddlesome) definitions of public health help spur state actions—including the regulation of product and labor markets—that in all likelihood jeopardize the health of the very individuals the new public health seeks to protect. The new public health extends regulation into inappropriate areas, and thus saps the social resources and focus to deal with public health matters more narrowly construed.

Epstein judged that there were at least three ways in which the new public health threatened to sap resources. First, deploying resources to the new public health would divert resources away from areas in which they could be employed more beneficially, especially contagious disease protection and provision of health care desired by individuals. Second, it could create counter-productive incentives. For example, Epstein thought that policies aimed at tobacco might undermine incentives for people to take charge of their own health by stopping smoking, instead perhaps leading them to believe they could recover for any harm they suffered from smoking by bringing damage suits against manufacturers or by receiving ameliorative health care at public expense. Here, Epstein argued that health insurers in the private market should take measures to prevent what is called "moral hazard": the tendency for people to engage in more of a risky activity if they believe they will receive compensation when things turn out badly and the risk falls on them. Governmental intervention, he contended, could instead further moral hazard. New public health regulations might also encourage manufacturers to take measures to avoid any regulatory impact, thus incurring expenditures without any value to consumers. Third, new public health measures could direct interventions against individuals in ways that would not be beneficial to them. Epstein used policies such as increasing taxes on sugary soft drinks to combat obesity as illustrations, arguing that they imposed costs on all consumers, whether or not they are at risk of obesity or find its consequences deleterious. These taxes may have a disparate impact on those who are less able to pay, moreover, actually reducing the abilities of low-income families to purchase nutritional food if they also wish to purchase soft drinks. With respect to the impact of new public health policies on individuals, Epstein invoked the familiar view in classical liberalism that individuals are the best judges of their own interests.

This libertarian assault on the new public health continues to be influential not only in U.S. public health policy but also in U.S. health policy more generally. States that are reluctant to adopt the Medicaid expansion of the Affordable Care Act continue to try to implement policies aimed at placing more responsibility on individuals, such as co-payments for care or requirements to work in order to receive benefits. So-called "right to try" laws that permit terminally ill individuals to receive experimental drugs from pharmaceutical companies, bypassing the approval process of the Food and Drug Administration, have swept the country, spurred by support from conservative organizations such as the Goldwater Institute and Cato Institute.

Responses to these attacks on the new public health include criticisms of their approach to legal history and constitutional law. Defenders of the new public health also argue that its libertarian critics confuse questions of population health with questions of individual health and mistakenly infer conclusions about populations from conclusions about individuals. Other defenders of the new public health take on its ethical position directly, contending that there are non-paternalistic reasons for many new public health initiatives and that paternalism is ethically justified in some situations. The remainder of this chapter takes up each of these lines of argument. A theme that runs through the analysis is that surveillance for the new public health may be justified even when direct interventions with individual behavior is not, as it generates information about social conditions and resource needs rather than coercing individuals to change their behavior. Some of the fissures in the U.S. health care system that are being revealed by COVID-19 lend further support to this argument.

6.4 U.S. Constitutional History, the New Public Health, and the Powers of Government

A threshold line of argument against Epstein's excoriation of the new public health was the critique of his argument that the new public health involved departures from settled traditions in U.S. constitutional law. U.S. constitutional law protects individual liberty and property rights; Epstein contended that the new public health trampled over these protections. His critics replied that the right to property as recognized in U.S constitutional law had never included rights to use property in ways that could cause serious and unavoidable harm to others. Both new and old public health, these critics said, aimed to reduce diseases across the population (e.g. Gostin and Bloche 2003). U.S. law had long recognized the importance of the state interest in protecting health. What had changed was the pattern of diseases causing mortality and morbidity within the population. Old and new had the same goal—health—but achieving it took different forms in different circumstances and required different agendas. Along these lines, Wendy Parmet (2009) developed a thorough and far-reaching account of how U.S. law has recognized and furthered efforts to pursue population health.

Epstein's libertarian conservatism rests on a very strong view of private property rights on which interference is justified only to protect others from what the law judges to be "nuisances." Nuisances are wrongs imposed on either the public or other owners through the unreasonable use of one's property in a way that substantially and unjustifiably interferes with the public or with private owners' enjoyment of their property. Someone's property may be a nuisance if it emits pollutants and someone's body may be a nuisance if it emits organisms causing infections in others. Nuisances may be private civil wrongs giving rise to tort liability if they harm individual property owners such as neighbors. They may be public nuisances and subject the owner to damage remedies, civil fines, or even criminal charges if they harm the public, as a rat-breeding home might do. Epstein's account of U.S. legal history contends that the courts construed the scope of nuisances narrowly until the twentieth century. His critics identify a wide range of nuisances recognized in legal doctrines from colonial times.

While a full account of the U.S. constitutional law of property rights and nuisance is far beyond the scope of this volume, a brief survey may provide a helpful background for libertarian legal challenges faced by public health in the U.S. Under the 5th and 14th Amendments to the U.S. constitution, the federal government and state governments respectively are prohibited from taking property without due process of law and just compensation. This means that if the government seizes an owner's property rights, it must have an adequate justification and it must pay for the privilege. So, for example, if an owner has a right to keep pigs on his property no matter the odors they might release, but the government steps in to stop the pig breeding because people nearby claim the odors are making them sick, the government must pay the owner compensation for the lost value in using his property as a pig farm. This analysis assumes, however, that the property owner had rights in the first place that the government has impinged. Another longstanding U.S. legal doctrine is that private property rights do not include the right to maintain an unwarranted interference with a public right; there is no property right in maintaining a public nuisance (*Mugler* 1887). Abating such a nuisance does not require compensation because the owner never had the right in the first place.

The scope of this doctrine of public nuisance is where controversy lies, however. In response to circumstances of the times, U.S. law has adjusted and re-adjusted what constitutes unreasonable uses of property that cause substantial harm to the public. *Mugler v. Kansas*, the leading Supreme Court case holding in 1887 that abatement of a public nuisance did not require compensation, involved prohibition of the manufacture of alcohol. Today, the public nuisance doctrine is being invoked by governments in the litigation against opioid manufacturers, although it is unclear whether this argument will ultimately succeed. Arguably, a rat-infested house is a clear case of a public nuisance justifying intervention. However, many cases are far less clear, especially when they involve regulations that affect property uses without the government's assumption of ownership. It is also an established doctrine in U.S. constitutional law that governments may not "go too far" in regulation limiting property uses (*Pennsylvania Coal* 1922). In working out the balance of what it is for regulation to "go too far" and thus amount to a regulatory taking, U.S. courts have shifted back and forth from stronger and weaker doctrines protecting property owners.

The primary area in which contemporary regulatory taking jurisprudence has been developed is not health but environmental protection, where building restrictions have been imposed to protect shorelines or wetlands. The currently controlling case, *Lucas v. South Carolina Coastal Council*, was decided in 1992. The owner of two lots on a barrier island was denied a building permit under coastal management legislation adopted after his purchase of the property. In an opinion written by Justice Scalia, the Court reaffirmed that any permanent physical invasion of property requires compensation, no matter how minimal the invasion and how weighty the public purpose. Presumably on this analysis, construction of a permanent station on private land for detecting harmful emission levels would require compensation for the invasion. For regulations "going too far" in affecting beneficial uses of the property, the Court held, the test is whether the proposed prohibited uses are beyond what is included in the owner's rights over the land in the first place. Applying this test requires balancing several factors: the degree of harm to the public or adjacent private property posed by the owner's proposed use, the social value of the owner's use and its suitability to the locality, and the relative ease by which the alleged harms from the use can be avoided (*Lucas* 1992, 1031). Landowners do not have rights to engage in activities that spill over to harm others or the public generally, so emission control regulations would be permissible. But if a use such as a grain storage facility has been longstanding, grain storage is a socially useful activity, and flea-carrying rats can be trapped as they enter adjoining neighborhoods, prohibitions of grain storage could be considered confiscatory and require compensation unless alternative beneficial uses of the property remain.

Some new public health efforts have encountered objections that they are regulatory takings, but the case law on this question is sparse. Massachusetts' law requiring disclosure of ingredients of tobacco products was held a regulatory taking with the court reasoning that it would have great economic impact on tobacco companies' reasonable investment-backed expectations in the formulas they used to make their products without sufficiently promoting public health (*Philip Morris* 2002). The property in question was trade secrets in the products' ingredients. The court's reasoning that the state's justification was insufficient was that there was only speculative evidence that disclosure of ingredients would affect consumer behavior. On the other hand, food safety regulations such as prohibition on egg sales from farms that have tested positive for salmonella have been held not to require compensation of the property owner (*Rose Acre Farms* 2004). Nor do health warnings about the possibility that perishable agricultural products may be contaminated by salmonella effect a taking, despite the owner's lost sales (*Dimare Fresh* 2015).

Cases invoking the public nuisance doctrine have been brought against tobacco companies, lead paint manufacturers, gun manufacturers, energy companies and, most recently, pharmaceutical companies making opioids. In these lawsuits, governmental entities have sought to recover public costs for health care, policing, public education, and other public services resulting from the harms they claimed were caused by the product in question. These are cases brought by the government seeking to recover costs for public health, not cases brought by private individuals seeking damages such as for cancers caused by smoking. The public case against tobacco companies settled (*Master Settlement Agreement* 1998), encouraging the suits that have followed. Other suits have not fared as well, however. States and municipalities have lost in some cases against paint manufacturers (*In re Lead Paint Litigation* 2007; *State* 2008) although California succeeded in claiming that manufacturers had promoted lead paint for indoor use until 1950 despite knowing it was harmful to children. (*People* 2017) Suits against energy companies claiming that the effects of fossil fuels on the climate are a public nuisance have failed (e.g. *Native Village of Kivalina* 2012). One case against gun manufacturers filed by the City of Gary, Indiana, in 1999 was still continuing as of early 2020 (*City of Gary* 2019), despite legislative grants of immunity to firearms companies for lawful sales of guns or harms resulting from third party wrongdoing. Public nuisance claims also are involved in the continuing litigation against opioid manufacturers (*In re National Prescription Opiate Litigation* 2020).

It is fair to say that legal doctrines about the scope of public health in the U.S. are continuing to evolve. Whether governments will be able to recover damages on the theory that health-harming products are public nuisances, or commercial entities will be able to assert property rights in their activities, will continue to be adjusted. The views of libertarian conservatives such as Epstein represent only one pole in the debates, but one that is nonetheless powerful, especially with the current make-up of the U.S. courts.

6.5   Populations or Individuals?

Other critics attack the libertarian position as confusing questions of individual health with questions of population health. These criticisms begin with the argument made by epidemiologist Geoffrey Rose (1985) that sick individuals and sick populations are fundamentally different problems requiring fundamentally different approaches. What explains the variance in individuals' susceptibility to disease—their genes, exposures, behaviors, or choices—may be quite different from what explains the variance in populations' susceptibility. Population-level approaches such as tobacco control, in Rose's view, would be far more effective in reducing consumption than efforts to modify individual behavior. This analysis would imply that while libertarians may be correct that regulation may affect particular individuals in ways that would not be beneficial to them, they are not correct that the same regulations would not be beneficial for public health overall. The Nuffield Council on Bioethics, in developing a stewardship framework for public health ethics, put the point thus:

> It takes only a moment's thought to recognise that many of the 'choices' that individuals make about their lifestyle are heavily constrained as a result of policies established by central and local government, by various industries as well as by various kinds of inequality in society. People's choice about what to eat, whether or not they allow their children to walk to school, or the kinds of products that are marketed to them, are often, in reality, limited. This means that the notion of individual choice determining health is too simplistic (2007, p. v).

Libertarians and their critics make quite different assumptions about the efficacy of population-level interventions and the role and capability of individuals, however. For libertarians, individuals are better judges of their own interests and have considerable ability to control their own lives. Population-level interferences may be wrong-headed on a grand scale. An example would be the assumption that health is a paramount value for the population when, for many members of the population, health is not at the top of their list. Many may judge health to be less valuable than other goods in their lives such as pleasures of consumption. For the libertarians' critics, by contrast, individuals may sometimes be wrong, or sometimes have information that is incomplete. Moreover, many features of individuals' health are beyond their control, no matter how hard they try or how carefully they choose. These features will require interventions at the population level. They go far beyond infectious disease or environmental toxins. No matter how much an individual might want fresh vegetables, she cannot will them into the local grocery store. Nor can she exercise successfully if the streets are unsafe or the air is polluted.

These points about the difference between population-level and individual-level questions hold for information, too. Some important information can only be gleaned at the population level just as effecting some changes in behavior require changes at the population level. Understanding the role of a wide variety of factors in causing disease, unearthing rare side effects or drug-drug interactions, or figuring out which gene variants are deleterious, cannot be accomplished with information about single individuals. In this respect, gathering information is like altering the built environment: just as individuals cannot will the presence of healthy food in their local stores, they also cannot will the information about rare events from a single case or even a significant group of individual cases. For this reason, it may be important for public health to have access to data about wide ranges of individuals.

The opioid problem illustrates how important population-level data may be. It took a considerable time after deaths from opioid overdoses started to rise for observers of health trends to recognize the extent of the problem. Evidence that rates of addiction to prescription opioids and opioid related deaths were rising began to appear over the first decade of the twenty-first century (Okie 2010). Opioid overdose deaths began rising sharply around 2000, when they surpassed rates of death from cocaine. A pub med search indicates that the earliest characterization of the situation as an "opioid crisis" or "epidemic" appeared about 2011 (e.g. Dhalla et al. 2011; Knoppert 2011; Manchikanti et al. 2012). In 2010, the FDA proposed a Risk Evaluation and Mitigation Strategy (REMS) to require physician and patient education about opioid risks. REMS was met with controversy because it was voluntary only. By that point in time, CDC data indicated that the highest rates of addiction and death were occurring in predominantly rural areas, although the data were incomplete because of variations in state surveillance systems. Had the data been better, the extent and shape of the problem might have been recognized earlier and addressed with greater success.

6.6 Paternalist and Non-paternalist Ethical Objections to the New Public Health

At the core of ethical objections to the new public health is paternalism. Described most generally, paternalism is interfering with someone's liberty for their own good when they do not agree to the interference (e.g. Dworkin 1972). So, it would be paternalistic to impose fines on willing smokers to get them to stop smoking and to prohibit people who are obese from purchasing sugary soft drinks. This highly general description elides important differences about what counts as interference (only coercion?), what liberties are at issue (to refuse knowledge?), what are the requirements for agreement (informed consent?), and when an interference is for someone's own good (what if someone agrees it was good after the fact?).

John Stuart Mill famously wrote in *On Liberty*:

> The object of this Essay is to assert one very simple principle, as entitled to govern absolutely the dealings of society with the individual in the way of compulsion and control, whether the means used be physical force in the form of legal penalties, or the moral coercion of public opinion. That principle is, that the sole end for which mankind are warranted, individually or collectively, in interfering with the liberty of action of any of their number, is self-protection. That the only purpose for which power can be rightfully exercised over any member of a civilised community, against his will, is to prevent harm to others. His own good, either physical or moral, is not a sufficient warrant. He cannot rightfully be compelled to do or forbear because it will be better for him to do so, because it will make him happier, because, in the opinions of others, to do so would be wise, or even right. These are good

reasons for remonstrating with him, or reasoning with him, or persuading him, or entreating him, but not for compelling him, or visiting him with any evil in case he do otherwise. (Mill 1859, pp. 17–18).

On Mill's view, paternalism unjustifiably thwarts liberty and does so in ways that are likely to prove counterproductive. Importantly, Mill applied his claims only to adults; whether paternalism may be justified in the case of children or "the race in its nonage" he regarded as separate questions.

Over the century and a half since Mill wrote, philosophers have debated whether his position rested fundamentally in his commitment to liberty or in his overall utilitarianism. The difference is this: if liberty is the foundation for anti-paternalism, then in the final analysis paternalism must be unjustified even if interference would have the best consequences overall. On the other hand, if utilitarianism is the foundation, then anti-paternalism must have the best consequences overall in order to be justified. If paternalistic interference would have better consequences overall, a utilitarian would have to admit that it is justified, so utilitarian anti-paternalists must contend as an empirical matter that in the long run interference would not work out for the best. And they have done so, claiming that interferers will misjudge what is good for others, that interference will be counterproductive, or that interference will be far too costly to outweigh any benefits. A typical way for utilitarian anti-paternalists to structure these empirical arguments is to apply them to rules: rules that prohibit paternalism will have the best consequences overall, they contend, even if their application to a particular case would appear counter-productive. So even if taxing sodas would prevent dire health consequences for a few, a policy of taxing sodas would burden others more or potentially backfire by causing resentment. Along similar lines, some might contend that restricting access to opioids, while it might do good in some cases, will result in more untreated pain overall.

Writers following Mill have also debated whether his commitment to liberty was to political liberty—the absence of state or social coercion—or to autonomy in a deeper sense of some kind of freedom of the will. In addition, what counts as coercion has been viewed very differently by commentators. Threats of injury or punishment are clearly coercive, but what about offers or changes in the structure of choices that are designed to change behavior? Behavioral economists call these changes in "choice architectures." Some argue that such an offer may coerce if it is so generous that a person has no choice but to accept it. An example would be paying very poor people substantial sums of money to lose weight. Some also argue that an offer may coerce if it would provide someone with a good that he or she has no right to receive. Paying kickbacks to clinics to allow patients to receive treatments for which they would not otherwise qualify would be an example.

Others, however, contend that offers do not coerce if they do not threaten to violate rights or obligations and people may choose to forego them, even if the choice might be to give up a benefit that is very attractive under the circumstances (e.g. Wertheimer and Miller 2008). So on this view, "wellness" discounts for health insurance offered to people who lose weight or stop smoking are not coercive, even if the discounts are very attractive, as long as people may choose not to accept them.

Some also argue that drug company payments to clinics for writing prescriptions would not coerce those receiving the prescriptions, although it might be objectionable on other grounds such as driving up the costs of health care.

Many less drastic changes in choice architecture alter available options or the ways in which these options appear to people in order to take advantage of insights from behavioral economics about how people make decisions. Changes in presentation of the items in a buffet line is the classic example: people will take more desserts if the desserts are prominently displayed than if they are on lower shelves or last in line. Wellness programs often use these devices, offering small rewards for behavioral changes such as better diet management or taking walks during lunchtime breaks. Whether such "nudges" are paternalistic and, if so, questionable interferences with liberty, is highly controversial in many areas of the new public health (e.g. Thaler and Sunstein 2008).

Ethical arguments for the new public health answer the charge of unjustified paternalism in a variety of ways. First, pursuing the goals of the new public health may not be paternalistic at all, if it occurs with consent. While it is paternalistic to make people do things for their own good, and perhaps even to force information on them against their wishes, it is not paternalistic to gather information that they actually consent to share. Whether it is paternalistic to gather information or to interfere with behavior without actual consent, but in ways people would have consented to, is more controversial, however. Second, there are non-paternalistic reasons for surveillance of NCDs, especially reasons rooted in justice; for example, data about the interplay between education and health may suggest needs for reform in primary and secondary education in the U.S. This is an argument from justice, not an argument from what would be best for individuals or even groups. Some contend, however, that as long as non-paternalistic reasons are coupled with paternalistic ones, problematic paternalism remains.

Finally, the paternalistic objection itself many be questioned. There are at least three lines of reply to the claim that new public health surveillance wrongly coerces people to do things for their own good. First, gathering information through surveillance does not entail intervening with behavior, although the information may lead to interventions. People may consent to learning about whether obesity and type 2 diabetes are connected without consenting to attempts to alter their weight; indeed, this information may be necessary for decisions to be well informed. Second, a distinction between impositions that override individuals' choices and impositions that result in what people would have chosen with adequate knowledge can be helpful. Finally, there are circumstances in which paternalism may be justifiable to avert significant harms, although these circumstances must be carefully delineated if surveillance is to be sustained.

6.7 Justifying Surveillance for the New Public Health without Paternalism

It is not paternalistic to interfere with someone who has voluntarily agreed to the interference. Of course, the devil is in the details about what counts as voluntary agreement. Nor it is paternalistic to interfere for reasons other than the individual's own good, such as for the protection of harm or injustice to others.

6.7.1 Agreeing to Give and Receive Information

Obtaining information when people consent to acquiring and receiving the information in order to make their own choices is not paternalistic. Information gathering is only paternalistic if it is designed to obtain information that people do not consent to provide or to receive, and to do so for their own good. Cholesterol testing is a good example. If a patient is required to have his cholesterol tested for his own good, and does not consent, this would be paternalistic. If he consents to the testing because he wants to know his own health status or wants to help other family members know whether there is a family history of high cholesterol, undergoing the test would not be paternalistic. It would, however, be paternalistic to require him to receive his results if he does not wish to have the information but is being tested only for the benefit of his family.

Surveillance for new public health objectives is not paternalistic if it obtains information that people actually consent to have collected and does not force them to receive information for their own good when they do not consent to do so. Much information obtained and used by the new public health would not violate these constraints. Individuals may willingly provide and receive health-related information. Problems occur when either some do not want to give information or some do not want to hear it.

An important complication here is the point made above about the difference between population-level and individual-level questions. The new public health uses population-level data, but this data must be gathered from individuals, even if they are not identified in how the information is gathered. Some would object that it is unjustified to collect information from individuals for the good of the population unless the individuals would agree to collection of the information for their own good. Flanigan (2013) for example argues that unless individuals consent to collection of information from them, it is paternalistic to coerce them to provide the information for the overall public good. Instead, she contends, they must be permitted to opt out from this kind of surveillance in order to respect their liberty. Cholesterol again may illustrate. Suppose that at the population level it is valuable to be able to answer questions such as: what are the correlations between cholesterol levels and health consequences? What characteristics of people or their behavior might affect these correlations—for example, do people with high cholesterol who exercise tend to have fewer adverse cardiovascular events than people who do not? What are the benefits and side effects of various anti-cholesterol medications? What characteristics of people or their behavior might be related to the frequency of these benefits or side effects? Meaningful answers to these questions will require data that are sufficiently representative of the population, especially to detect low-frequency events; allowing people to opt out may compromise the analysis. The question is whether collecting population level data for the good of the population is paternalistic unless the collection is modeled on individual consent, as Flanigan argues that it is.

Arguably, collecting the data to answer these questions is not paternalistic because it is not done for any individual's good against his own choice. Instead, it is done at the population level for the benefit of the population. To be sure, there may be individuals who would prefer not to participate in the gathering of population-level data for many different reasons and would not consent to have data drawn from them used in the effort to answer any of these questions. Whether it is justified to use data from them, however, is not a question of paternalism; it is a question of the ethics of using data drawn from individuals who do not and would not consent, when the data are needed for the benefit of others. Whether data are needed depends on whether any opt out policies would undermine representativeness—itself difficult to know without having at least some data about the nature of the population. Privacy scholar Mark Rothstein (2010) has argued that de-identification is insufficient to protect privacy because of risks of stigmatization and data uses individuals would find objectionable. He notes the possibility of selection bias in data but contends that this speculative risk must be balanced against other potential harms. Others have noted that selection bias may be particularly problematic for underrepresented groups (e.g. Hoffman 2010). These groups may be small in number in the first place and may also for historical, religious, or cultural reasons be more likely to opt out. Such opting out will potentially compromise the information available to group members who do not wish to opt out, so the argument is that the data collection is not paternalistic but for the benefit of these others. Justification must, then, rest on other grounds such as the equity considerations explored in Chapter 4.

Groups present complicated issues in this regard, however. Some members of a group may disagree with the choices of others about the good of the group and argue that it is unjustified paternalism to make them share information for the good of the group. The population-level reply given in the preceding paragraph would be that this is not paternalism because the individuals are being required to share information for the good of others. However, it is possible for groups or even nations to be the subject of paternalism. Michael Barnett (2015) argues that forms of global governance such as those encouraged by the WHO are paternalistic if they impose humanitarian goals without the agreement of those to whom they are applied.

Suppose, for example, that members of a religious or ethnic group have immigrated and settled together in a new area. They are sufficiently cohesive to be described as a group; although they do not have a governing structure, some worship together, many join together for celebrations at a community center, and there are interlocking structures of friendship and familial relationships among members of the group. They retain dietary preferences from their homeland that have been associated with weight gain and high rates of diabetes among similar immigrant groups. Public health officials in the jurisdiction in which many of the group members reside note that there are high rates of obesity in the schools and of admissions for cardiovascular disease in local hospitals. Death certificates also indicate higher than average rates of premature death. Public health might therefore wish to gain information about the group and its dietary practices in order to determine whether or not further intervention might be warranted for the good of the group's health. Many group members, however, might object to the collection of information about their health, such as information from their medical records, grocery store purchases, and the like. This surveillance might be characterized as paternalism aimed at the group—information to enable analysis about health interventions that might be needed to protect the group against itself. (Surveillance likely would also violate equal protection if the information was collected from group members only because of their religion or ethnicity.). Even if recognizable group leaders consented to the surveillance on behalf of the group, it still might be characterized as paternalism against individual group members who object; we take up these questions about the implications of group consent in Chapter 7. Analogous points might be made about whether democratically made decisions to paternalize reflect the "consent of the governed" and therefore are not paternalistic to those who voted against them (Nys 2008), arguments that we also consider in Chapter 7. In the remainder of this chapter, we consider whether non-paternalistic arguments such as equity may be convincing, and whether, if not, there may be circumstances in which paternalism may be justified, whether it is directed against individuals or against groups.

6.7.2 Non-paternalistic Reasons for the New Public Health: Education and Social Determinants of Health

Paternalism means imposing choices on people for their own good. Many apparently paternalistic public policies also have non-paternalistic justifications. Arguably, these policies are therefore not paternalistic, although some would still argue that to the extent that the paternalistic justification for these policies carries weight, they remain problematic. The requirement to wear seat belts in automobiles or to wear motorcycle helmets may be justified to reduce the costs of accidents to others. Like wellness programs, however, some of these programs may primarily be efforts to protect people from themselves; if so, they are paternalistic. Some justifications for new public health surveillance, however, are clearly not paternalistic but are rooted in justifications such as equity or justice. An equity-based justification does not make people do things for their own good; it intervenes to eliminate unfair disparities that adversely affect marginalized groups, as we explained in Chapter 4.

Many of the efforts of the new public health are aimed at improving health equity. Take, as an example, evidence about education as a social determinant of health. Longstanding assumptions in U.S. political ideology are the role of individual choices in success and the need for education to give people opportunities to make these choices. One of the earliest lines of argument about the social determinants of health was that improved education would yield improved health. The first of the "Healthy People" assessments of the health of the U.S. population and how it might be improved, issued in 1979, attributed at least half of mortality in the U.S. to "unhealthy behavior or lifestyle" and only 10% to inadequate health care (DHHS 1979, 1–9). The report opined that "the health of this Nation's citizens can be significantly improved through actions individuals can take themselves, and through actions decision makers in the public and private sector can take to promote a safer and healthier environment for all Americans at home, at work, and at play" (DHHS 1979, 1–12). To enhance children's health, the report recommended increased support for preschool programs such as Head Start; it also emphasized the "special importance of the school" and how "[o]ur children could benefit greatly from a basic understanding of the human body and its functioning, needs, and potential— and from an understanding of what really is involved in health and disease" (DHHS 1979, 4–16). Education was judged core to reducing risks of teen pregnancy, smoking, poor diet, and drug and alcohol abuse. Studies published during the decade following in the *American Journal of Public Health* addressed topics such as the need to write smoking cessation manuals at a 5th grade reading level (Meade et al. 1989), correlations between smoking rates and education (Novotny et al. 1988), relationships between a wife's education level and her husband's susceptibility to coronary disease (Strogatz et al. 1988), and correlations between education and teen pregnancy (Joyce 1988). Editorials also weighed in, on topics such as the role of education in disparate mortality rates from cardiovascular disease between whites and blacks (Wing 1988) or poor nutrition (Editorial 1988). Improving education, it seemed, might be the very best way to improve population health.

The relationship between education and health is far more complex than any simple correlation between increasing education and improving health, however. Developing understanding of the interrelationship among social determinants of health and policy interventions that may prove successful in addressing health inequities requires ongoing data analysis. Here are several examples. In 2002, a review of socioeconomic status and health judged that environmental risk exposures were a major determinant of health and included school facilities among environmental factors. This study noted that low income schools were significantly more likely to have problems with building repair and infrastructure such as heating (10% vs. 30%) and to be overcrowded (6% verses 12%) and significantly less likely to have teachers with degrees in math or science teaching these subjects (27% versus 43%) (Evans and Kantrowitz 2002).

To take a more recent example, in a report prepared for the U.S. Agency for Healthcare Research and Quality, researchers reviewed the extensive evidence of the impact of education on individual, community, and population health (Zimmerman et al. 2015) and developed an ecological framework for analysis. The reported differences are significant: a 9-year difference in life expectancy between adults age 25 with a high school diploma and those with a college degree. Understanding what is at work behind these data about education and health is critical to designing policies to address them. For this, a far more complex evidentiary picture is needed. Zimmerman and colleagues sketch out what some of this evidence might be. There are many direct and indirect effects of education. Health literacy is one direct effect: people with higher levels of education are better able to understand their health needs, communicate with providers, understand care plans, and navigate the health care system. Adults with more education also can access other benefits: employment, income, and social networks and support. Education also correlates with reduced levels of risky behavior such as smoking, alcohol use, poor diet, and inactivity. Adults with more education are also at lower risk of long-term stress. Evidence also reveals neighborhood-level effects from factors such as the availability of resources, social capital, and social organization. Lower-income neighborhoods are more likely to be unsafe, have poor access to nutritious food and green space, and have poorer schools. The evidence suggests that the relationship between education, social contexts, and education is complex and interactive, however, and may be influenced by external factors such as legislative funding decisions and social safety net buffers. Nor can educational inequalities be separated from inequalities by gender, race, ethnicity, sexual orientation and disability, say Zimmerman and coauthors. Causality may go in reverse as well, if ill health adversely affects educational success because of increased absenteeism or cognitive difficulties. Education may also be a proxy for other factors, such as the lack of family stability. The authors recommend further research to better understand the connections between education and these other social factors, including research following the model of community-based research in which community members help to shape research questions.

An even more recent example questions the role of education, arguing that wealth is far more significant than education in affecting later educational outcomes and SES. According to this study, education and SES in turn will correlate with health, as just described. This study examined correlations between early measures of academic ability and educational success and found that half of the children from low socioeconomic status families who had high test scores in kindergarten were behind by eighth grade; for Black students from low SES families, the rate was even higher at 60% (Carnivale et al. 2019) and for Asian students it was the lowest, at 20%. The authors of this study concluded that in addition to good schools backed by sound educational policy, success requires providing students with the environmental supports they need for success.

These studies, and the need for the data to conduct them, reflect the strong equity concerns surfaced in Chapter 4. They involve children and likely opportunities. The disadvantages they involve are likely corrosive, interacting to yield even worse outcomes for the futures of those they affect. Surveillance to examine relationships among factors affecting opportunity is justified as a matter of social justice, not as a matter of protecting people for their own good.

On the other hand, some examples of surveillance for the new public health are paternalistic to either individuals or groups, even if they are described in other terms. Consider wellness programs that require individuals to "know their numbers" about blood pressure and cholesterol. These programs may be justified on non-paternalistic grounds such as reducing overall health care costs or workforce absenteeism, or perhaps selecting healthier employees or encouraging them to join or stay in the workforce (Jones et al. 2018). But these reasons may be thinly veiled paternalistic efforts to prod people to take better care of themselves and to take charge of their health. Indeed, some data suggest that while the programs do affect employee behavior they have no significant impact on absenteeism or health care costs (Song and Baicker 2019). If these programs require health-related information to be obtained and presented to individuals for their own good, whether or not the individuals want the information, they are paternalistic.

6.7.3 Non-paternalistic Arguments for the New Public Health: Public "Bads."

Economists and others who rely on economic reasoning such as Epstein point to market failures as justifying governmental intervention. Markets fail in the case of public goods because individuals can receive their benefits without incurring any of the costs of producing them, so have no incentive to pay the costs. Markets may also fail when private behavior spills over to impose costs on others as with pollution; in such cases, if the transaction costs of organizing individuals to oppose the pollution are high, there is a utilitarian economic case for intervention to force the polluter to internalize the costs of these externalities. Public health law professor Lindsay Wiley (2012, p. 213) and others argue that there are also public "bads"—negative externalities inflicted on the public without their consent—that should be the subject of governmental attention. Examples are environmental factors that encourage unhealthy eating or unsafe parks that make outdoor exercise very difficult. To defend the role of the state in addressing these public bads, Wiley points to the idea of a "public nuisance" described above: an unreasonable interference with a right common to the general public, such as a stream (2012, p. 235). In public nuisance suits, the required proof is of harm to the public in general, not to individual members of the public. The harms are to the collective rather than to individuals considered separately from one another, even a reasonably large number of them. Wiley's argument is that the science of epidemiology is yielding information about how the social environment affects population health, information such as that summarized in the discussion of education above. To refuse to expand the scope of public health to these public bads, Wiley argues, is to detach practice from science. On this view, epidemiological harms such as the food or transportation environment should be addressed by public health, just as individual harms such as salmonella in a bunch of spinach consumed by a particular purchaser should be.

For many public policies there will be both paternalistic and non-paternalistic reasons that can support them; these might be called "mixed motive" paternalism. Sometimes, the non-paternalistic reasons offered for a policy are thinly veiled excuses for paternalism. Moreover, critics might argue (e.g. Flanigan 2013), the objections to paternalism are so strong that the non-paternalist reasons are overridden by them. So, the question of paternalistic justifications for the new public health cannot be avoided. We now take on the question of paternalism directly.

6.8 Paternalist Arguments for the New Public Health

Is it ever justifiable to make people do things for their own good? Is paternalism more plausible depending on the importance of the good? If so, is health an especially compelling good? Does the justification of paternalism change depending on whether coercion is used, or whether softer methods can achieve the same ends?

6.8.1 Justifying Paternalism

Contemporary debates about paternalism take the form of asking whether paternalism can be justified. They start out with the assumption that it is paternalism—not the opposite—that requires justification. It is worthwhile pointing out, however, that this way of structuring the argument itself makes the assumption that the burden of justification must be borne by the paternalist. That this assumption seems so reasonable stems from presumptions about the importance of liberty: that it is interferences with liberty that require justification, not liberty itself. The normative burden of persuasion is on the defender of interference, not the defender of liberty. Perhaps liberty is so critical a value that it is plausible to place the burden of persuasion on those who would limit it. However, the problem of justifying paternalism might be conceptualized quite differently in views that were not wedded to liberalism by prioritizing liberty in this way.

The presumption of liberty just described functions in an argument against interference with individuals' choices about how to live their lives. It thus applies to questions such as whether to interfere with people's food choices in order to promote healthy weight. Applying these presumptions to surveillance rather than to choices about health-related behavior, however, may be less plausible. Take, for example, surveillance about the effects of soft drink consumption on health. Is the evidentiary burden on public health to show that consumption of soft drinks is associated with weight gain that is in turn associated with diabetes, or on opponents of regulation to bring evidence that the associations do not exist? Is the evidentiary burden on public health to show that there are no unanticipated consequences of soft drink regulation, or is the burden on opponents to show that there are such unanticipated consequences? The logic is similar with opioids: does the proponent of intervention bear the burdens of demonstrating harm or the non-existence of unanticipated effects? The answers may have significant consequences in practice, if interventions are delayed for considerable periods in order to allow time to gather and assess the evidence.

Evidentiary burdens of persuasion have implications for the need for surveillance. If evidence is required to justify paternalism, but cannot be obtained, the result will be inability to justify the paternalism. Strategies for keeping evidence at bay will thus prove successful in keeping intervention without apparent justification—as tobacco companies and firearms manufacturers have realized to their advantage. Thus, there may be reasons to separate surveillance questions from paternalism: having the information may be critical to deciding whether an intervention is paternalistic or, even if so, whether it can be justified. In the words of Chokshi and Stine (2013), a "savvy" state is different from a "nanny" state.

6.8.2   Justifications for "Softer" Paternalism

The term "soft" paternalism was initially coined to refer to forms of interference that are needed to ascertain whether an individual is acting voluntarily (e.g. Dworkin 2017). Other writers have used the term to describe interferences that are less invasive than the use of outright prohibitions or serious threats (e.g. Conly 2013; Nys 2008). In this section, we consider the justification of "softer" forms of paternalism of these two types. Analogously, political scientists use the term "soft power" to describe the ability of a country to persuade others do to what it wants through shaping long-term attitudes and preferences (Nye 2004).

One kind of soft paternalism involves providing people with information in order to be sure they are acting voluntarily. Sometimes bystanders may be aware that others are in danger, but not know whether their danger is willingly incurred. Is it paternalistic to intervene in such cases? Mill himself argued that it would be justified to stop someone from crossing an unsafe bridge who did not know of the bridge's condition because this would not interfere with his liberty (1859, pp. 182–183). But for Mill this justification extends only to providing the information: "Nevertheless, when there is not a certainty, but only a danger of mischief, no one but the person himself can judge of the sufficiency of the motive which may prompt him to incur the risk: in this case, therefore (unless he is a child, or delirious, or in some state of excitement or absorption incompatible with the full use of the reflecting faculty), he ought, I conceive, to be only warned of the danger; not forcibly prevented from exposing himself to it." Surveillance needed to provide information would not be problematic soft paternalism, on this view, but it would be problematic to go further in attempting to change behavior.

Lack of information is not the only way in which behavior might not be fully voluntary. Behavioral economists Richard Thaler and Cass Sunstein have popularized the idea of "nudges" as a form of paternalism which, they argue, should be acceptable to libertarians. Nudges work with the observation that human decisions are beset with cognitive biases: we respond to the last event as most salient, are less willing to incur losses than to give up similar gains, and place disproportionate weight on the present. Nudges are defended as responses to the irrationalities created by these biases. For example, the bias in favor of eating a large piece of chocolate cake in the present might be countered by a mirror behind the food line that shows everyone five pounds heavier and information about the calorie count of the piece of cake.

Even if nudges take advantage of biases rather than countering them, some argue, they are permissible as long as people are made aware of them (Aggarwal et al. 2014; Thaler and Sunstein 2008). Someone is not really tricked into foregoing the chocolate cake by an image of her with five extra pounds, on this view, as long as she is aware that she is being shown the image in order to discourage her from taking the cake. This line of argument—that the person is not tricked—does not show that the interference is not paternalist, however. The interference is still for the person's own good, even though it is designed to get her to think about the future rather than responding to the immediate present. Rather, it shows at best that the interference may not be particularly burdensome; with awareness that the interference is occurring, people may take the cake anyway.

Some theorists defend interferences that may be more burdensome. Sarah Conly (2013) argues that paternalism may be justified to help people pursue their long-term goals, even though in the short run it prohibits them from having what they want. Indeed, Conly contends that prohibitions may actually be more respectful of people as choosers than "softer" measures such as information or nudges, because they are more effective in helping people to achieve their long-term goals. In response to the objection that paternalism treats people unequally because it assumes that some know better than others what is good for them, she counters that paternalism simply recognizes that rationality is flawed for everyone. This recognition, she argues, does not fail to respect us as persons; rather, it assesses our abilities accurately and values our longer-term choices.

Along similar lines, Thomas Nys (2008) argues that "deeper" autonomy justifies paternalism in public health. He assumes the value pluralism of liberalism—that paternalism cannot be justified on the basis that it is objectively true that some things are goods for everyone. Instead, people may value different things, or value the same things differently; some may value health or even life more than others do. However, people sometimes make choices that go against or jeopardize their fundamental conceptions of their good. On Nys' view, interferences that are minimal are permissible; he uses information gathering through a cheek swab as an illustration of an intervention that does not interfere with broader autonomy about major life choices. Nor does it interfere with deep autonomy in the sense of being reflective about the worth of these major life choices.

These defenses of softer paternalism are liberal in the sense that they assume value pluralism. They accept that people may hold a range of different values and may balance these values differently. Longer life may simply not be a value for some. Or, while people may share the value that death is bad, they may balance it differently against other goods. Some—Kneiss (2015) calls them "food lovers"—may value the enjoyment of fatty foods more than the enjoyment of additional days or even years of life. Soft paternalists do not undo this evaluative structure; rather, they intervene to further people on the way of realizing this value structure. The intervention is justified in terms of helping people to realize their goals, not in terms of helping them to have better goals. Often this justification is put in terms of a balance, with the intervention seen as most justified when it is minimal but helps people achieve things that are important to them on their own terms. In the end this line of reasoning must yield when individuals insist that behavior is in accord with their ultimate values. Importantly, however, the line of reasoning has greater purchase when the interference is limited to information that will help individuals to determine whether the behavior really is in accord with their ultimate values. Unless an individual has grounds to argue that acquiring the information will compromise her other ultimate values in some way, or that it is against her ultimate values to have the information she needs to determine what is in her ultimate values, surveillance would not be impermissible on this line of soft paternalist reasoning.

6.8.3 Combining Soft Paternalism with Fairness to Others

Kneiss (2015) argues that paternalistic policies may be justified for reasons of fairness combined with soft paternalism. Some behavior is not entirely self-regarding; second-hand smoke from tobacco is an illustration. Sometimes, behavior is not entirely self-regarding because of the impact it has on how others may behave. Suppose some people are poorly educated about food or particularly prone to cognitive bias. Perhaps they are especially vulnerable to the influences of others. Suppose that regulations about food cannot be easily tailored to protect these people while food lovers who do not care about their health continue to eat as they wish. If so, Kneiss argues, it may be unfair to those who are vulnerable not to intervene paternalistically. He considers the objection that such vulnerability may be due to social conditions such as social determinants of health discussed earlier in this chapter and that, if so, it would be preferable to address the social conditions rather than intervening paternalistically. His reasoning is rooted in non-ideal theory: that in circumstances where social conditions are inequitable, addressing factors related to ill health directly may be part of an effective strategy for addressing unjust social conditions. By reducing ill health that leads to inequality of opportunity, we may address social disparities, rather than the reverse. To hold otherwise, Kneiss concludes, is to privilege liberty over opportunity.

6.8.4 Justifying Hard Paternalism?

"Hard" paternalists believe that it is sometimes justifiable to intervene to make sure that people have what is good for them. Hard paternalism requires a justification for the importance of the good and reasons for thinking that the good is more important than any costs of interference. Many hard paternalists are perfectionists rather than liberals: they believe that there are some goods that are so important for human life that everyone ought to have them, whatever their own views about their good.

Examples might be life itself or physical health. These positions are illiberal in that they fail to recognize individuals as ultimate sources of their good. Resolving fundamental questions of political philosophy about liberalism is beyond the scope of this volume. Suffice it to say at this point of the argument that if the justification for surveillance is ultimately rooted in this form of hard paternalism, it must reject liberal commitments to individuals as ultimate sources of the good for themselves.

Others take a somewhat less perfectionist balancing approach, weighing the importance of the choices to people, the degree of the interference, and the good to be achieved by interference. Wilson (2009), for example, contends that interests such as not wearing a seat belt are simply not very important to people. Some take this position about providing information. Nys (2008), for example, claims that providing information is not a particularly serious intervention because it is only temporary and does not interfere with the ultimate exercise of autonomy. The Nuffield Council (2007, xix), in its analysis of the justification of public health interventions, structures an "intervention ladder"; monitoring the situation and providing information are the lowest rungs on the ladder. They place collecting anonymous information about outbreaks of contagious disease low on the ladder. Case reports are higher because of the potential for intervention, however. If these positions are based in the view that privacy in the form of protection from being required to share information about oneself is sufficiently unimportant for everyone that it may be outweighed by other goods, they, too, must reject liberalism's picture of individuals as ultimate sources of their good.

Wearing masks during the COVID-19 pandemic has become a particularly contentious issue in the U.S. It is an excellent example of these issues about paternalism. The non-paternalist argument for mask-wearing is that it protects others than the mask-wearer from the possibility of infection. But there are paternalistic arguments for mask-wearing, too, of the kinds sketched above. Wearing a mask protects the wearer. The wearer may not be a good judge of whether she is infected or likely to be at risk for others. The person chafing at mask-wearing may irrationally discount the data about the benefits of masks or may erroneously believe that masks deprive the wearer of oxygen. The mask-rejecter may confuse a momentary and minor inconvenience for an important value. People who reject any hint of paternalism as support for mask-wearing may contend that their rejection of paternalism is so important that it outweighs the non-paternalistic argument that masks protect others. In this vein, some who want not to wear masks have stated outright that they believe they have the right to place others at risk. For example, a Utah woman opposed to wearing masks stated, "I don't think it's the government's place to tell me I should or shouldn't wear a mask. And if I want to take that risk or if I choose to put people at risk, that should be up to me" (Walker 2020).

6.9 Summary

Straight-out forms of hard paternalism must reject liberalism. If hard paternalism is required to defend surveillance, it must ultimately substitute judgments about their good for people's own determinations. However, hard paternalism may not be required to support much surveillance. Non-paternalistic justifications are available, such as consent, prevention of public nuisances, and equity. Softer paternalistic justifications are also available, such as exploring how providing people with information may help them to realize their own values in the longer term. These softer informational considerations may intertwine with non-paternalistic justifications such as fairness to others who need the information to pursue their conceptions of their good. Some who object to hard paternalism have insisted on informed consent for individuals to be required to share information about themselves for purposes of public health surveillance. In the next chapter, we explore how liberal values of respect for individuals as sources of their own good can be reflected in surveillance even without relying on models of individual informed consent.

References

Aggarwal, Ajay, Joanna Davies, and Richard Sullivan. 2014. "Nudge" in the Clinical Consultation—An Acceptable Form of Medical Paternalism? *BMC Medical Ethics* 15: 31.

Barnett, Michael. 2015. Paternalism and Global Governance. *Social Philosophy & Policy* 32: 216–243.

Campion, Edward W. 2018. The Problem for Children in America. *New England Journal of Medicine* 379: 2466–2467.

Carnivale, Anthony P., Megan L. Fasules, Michael C. Quinn, and Kathryn Peltier Campbell. 2019. *Born To Win, Schooled to Lose: Why Equally Talented Students Don't Get Equal Chances to Be All They Can Be.* Georgetown University Center for Education and the Workforce. https://1gyhoq479ufd3yna29x7ubjn-wpengine.netdna-ssl.com/wp-content/uploads/FR-Born_to_win-schooled_to_lose.pdf. Accessed 1 Aug 2020.

Case, Anne, and Angus Deaton. 2015. Rising Morbidity and Mortality in Midlife Among White Non-Hispanic Americans in the 21st Century. *Proceedings of the National Academy of Sciences* 112 (49): 15078–15083.

Centers for Disease Control (CDC). 2020. *National Diabetes Statistics Report, 2020.* https://www.cdc.gov/diabetes/pdfs/data/statistics/national-diabetes-statistics-report.pdf. Accessed 1 Aug 2020.

Chokshi, Dave A., and Nicholas W. Stine. 2013. Reconsidering the Politics of Public Health. *Journal of the American Medical Association* 310 (10): 1025–1026.

City of Gary v. Smith & Wesson Corporation, 126 N.W.3d 813 (Ind. App. 2019).

Conly, Sarah. 2013. Coercive Paternalism in Health Care: Against Freedom of Choice. *Public Health Ethics* 6 (3): 241–245.

Cunningham, Rebecca M., Maureen A. Walton, and Patrick M. Carter. 2018. The Major Causes of Death in Children and Adolescents in the United States. *New England Journal of Medicine* 379: 2468–2475.

Department of Health and Human Services (DHHS). 1979. *Healthy People: The Surgeon General's Report on Health Promotion And Disease Prevention.* https://profiles.nlm.nih.gov/ps/access/NNBBGK.pdf. Accessed 31 July 2020.

———. 2019. *What is the U.S. Opioid Epidemic?* https://www.hhs.gov/opioids/about-the-epidemic/index.html. Accessed 31 July 2020.

Department of Public Health. 2018. *Reportable Diseases and Conditions*. https://www.cdph.ca.gov/Programs/CID/DCDC/Pages/Reportable-Disease-and-Conditions.aspx. Accessed 31 July 2020.

Dhalla, Irfan A., Navindra Persaud, and David N. Juurlink. 2011. Facing Up to the Prescription Opioid Crisis. *British Medical Journal* 343: d5142.

Dimare Fresh, Inc. v. U.S., 808 F.3d 1301 (Fed. Cir. 2015).

Dworkin, Gerald. 1972. Paternalism. *The Monist* 56 (1): 64–84.

———. 2017. Paternalism. *The Stanford Encyclopedia of Philosophy* (Winter 2017 Edition), Edward N. Zalta ed. https://plato.stanford.edu/archives/win2017/entries/paternalism/. Accessed 31 July 2020.

Editorial. 1988. The Nutrition Connection: Why Doesn't the Public Know? *American Journal of Public Health* 78 (9): 1147–1148.

El, Bcheraoui, Ali H. Mokdad Charbel, and Laura Dwyer-Lindgren. 2018. Trends and Patterns of Differences in Infectious Disease Mortality Among US Counties, 1980–2104. *Journal of the American Medical Association* 319: 1248–1260.

Epstein, Richard A. 2003. Let the Shoemaker Stick to His Last: A Defense of the "Old" Public Health. *Perspectives in Biology and Medicine* 46 (3 Suppl): S138–S159.

Ethridge, Lynn. 1983. Regan, Congress, and Health Spending. *Health Affairs* 2 (1): 14–24.

Evans, Gary W., and Elyse Kantrowitz. 2002. Socioeconomic Status and Health: The Potential Role of Environmental Risk Exposure. *Annual Review of Public Health* 23: 303–331.

Flanigan, Jessica. 2013. Public Bioethics. *Public Health Ethics* 6 (2): 170–184.

Gostin, Lawrence O., and Maxwell Gregg Bloche. 2003. The Politics of Public Health: A Response to Epstein. *Perspectives in Biology & Medicine* 46.e (Suppl): S160–S175.

Hanson, Victoria, Eyal Oren, Leslie K. Dennis, and Heidi E. Brown. 2016. Infectious Disease Mortality Trends in the United States, 1980–2014. *Journal of the American Medical Association* 316 (20): 2149–2151.

Hoffman, Sharona. 2010. Electronic Health Records and Research: Privacy Versus Scientific Priorities. *American Journal of Bioethics* 10 (9): 19–20.

In re Lead Paint Litigation, 924 A.2d 484 (N.J. 2007).

In re National Prescription Opiate Litigation, 2020 WL 582151 (United States Judicial Panel on Multidistrict Litigation 2020).

Jack, Shane P.D., Emiko Petrosky, Bridget H. Lyons, Janet M. Blair, Allison M. Ertl, Kameron J. Sheats, and Carter J. Best. 2018. Surveillance for Viodent Deaths—National Violent Death Reporting System, 27 States, 2015. *MMWR—Surveillance Summaries* 67 (2): 1–32.

Jacobson, Peter D. 2014. Changing the Culture of Health: One Public Health Misstep at a Time. *Society* 51: 2210228.

Jones, Damon, David Molitor, & Julian Reif. 2018. *What Do Workplace Wellness Programs Do? Evidence from the Illinois Workplace Wellness Study*. NBER Working Paper No. 24229. https://www.nber.org/papers/w24229. Accessed 31 July 2020.

Joyce, Theodore. 1988. The Social and Economic Correlated of Pregnancy Resolution among Adolescents in New York City, by Race and Ethnicity: A Multivariate Analysis. *American Journal of Public Health* 78 (6): 626–631.

Kneiss, Johannes. 2015. Obesity, Paternalism and Fairness. *Journal of Medical Ethics* 41: 889–892.

Knoppert, David. 2011. The Worldwide Opioid Epidemic: Implications for Treatment and Research in Pregnancy and the Newborn. *Pediatric Drugs* 13 (5): 277–279.

Lucas v. South Carolina Coastal Council, 505 U.S. 1003 (1992).

Master Settlement Agreement. 1998. https://publichealthlawcenter.org/sites/default/files/resources/master-settlement-agreement.pdf. Accessed 31 July 2020.

Manchikanti, Laxmaiah, Standiford Helm II, Bert Fellows, Jeffrey W. Janata, Vidyasagar Pampati, Jay S. Grider, and Mark V. Boswell. 2012. Opioid Epidemic in the United States. *Pain Physician* 15: ES9–ES38.

Meade, Cathy D., James C. Byrd, and Martha Lee. 1989. Improving Patient Comprehension of Literature on Smoking. *American Journal of Public Health* 79 (10): 1411–1412.

Mill, John Stuart. 1859. *On Liberty*, many eds. Available from Project Gutenbert ebook#34901. http://www.gutenberg.org/files/34901/34901-h/34901-h.htm. Accessed 31 July 2020.

Mugler v. Kansas, 123 U.S. 623 (1887).

Native Village of Kivalina v. ExxonMobil Corporation, 696 F.3d 849 (9th Cir. 2012).

Ncayiyana Daniel J. 1995. The New Public Health and WHO's Ninth General Programme of work: A Discussion Paper. Geneva: World Health Organization; 1995. https://apps.who.int/iris/bitstream/handle/10665/63061/WHO_HRH_96.4.pdf?sequence=1. Accessed 31 July 2020.

Novotny, Thomas E., Kenneth E. Warner, Juliette S. Kendrick, and Patrick L. Remington. 1988. Smoking by Blacks and Whites: Socioeconomic and Demographic Differences. *American Journal of Public Health* 78 (9): 1187–1189.

Nuffield Council on Bioethics. 2007. Public Health: Ethical Issues. Cambridge, UK: Cambridge Publishers Ltd. http://nuffieldbioethics.org/wp-content/uploads/2014/07/Public-health-ethical-issues.pdf. Accessed 31 July 2020.

Nye, Joseph S., Jr. 2004. *Soft Power: The Means to Success in World Politics*. New York: Public Affairs.

Nys, Thomas R.V. 2008. Paternalism in Public Health Care. *Public Health Ethics* 1 (1): 64–72.

Office of Disease Prevention and Health Promotion (ODPHP). 2020. *Healthy People 2020 Leading Health Indicators: Progress Update*. https://www.healthypeople.gov/2020/leading-health-indicators/Healthy-People-2020-Leading-Health-Indicators%3A-Progress-Update. Accessed 31 July 2020.

Okie, Susan. 2010. A Flood of Opioids, a Rising Tide of Deaths. *New England Journal of Medicine* 363: 1981–1985.

Parmet, Wendy E. 2009. *Populations, Public Health, and the Law*. Washington, DC: Georgetown University Press.

Pennsylvania Coal Co. v. Mahon, 260 U.S. 393 (1922).

People v. ConAgra Grocery Products Co., 227 Cal. Rptr.3d 499 (Cal. App., 6th Dist. 2017).

Philip Morris, Inc. v. Reilly, 312 F.3d 24 (1st Cir. 2002).

Rose, Geoffrey. 1985. Sick Individuals and Sick Populations. *International Journal of Epidemiology* 14 (1): 32–38.

Rose Acre Farms, Inc. v. United States, 373 F.3d 1177 (Fed. Cir. 2004).

Rothstein, Mark A. 2010. Is Deidentification Sufficient to Protect Health Privacy in Research? *American Journal of Bioethics* 10 (9): 3–11.

Song, Zirui, and Katherine Baicker. 2019. Effect of a Workplace Wellness Program on Employee Health and Economic Outcomes: A Randomized Clinical Trial. *Journal of the American Medical Association* 321 (15): 1491–1501.

State v. Lead Industries Association, Inc., 951 A.2d 428 (R.I. 2008).

Strogatz, David S., David S. Siscovick, Noel S. Weiss, and Gad Rennert. 1988. Wife's Level of Education and Husband's Risk of Primary Cardiac Arrest. *American Journal of Public Health* 78 (11): 1491–1493.

Thaler, Richard H., and Cass Sunstein. 2008. *Nudge: Improving Decisions About Health, Wealth, and Happiness*. New Haven: Yale University Press.

Tulchinsky, Theodore, and Elena Varavikova. 2001, 2d ed. 2008. *The New Public Health: An Introduction for the 21st Century*. San Diego: Academic Press.

Tulchinsky, Theodore, and Elena Varivikova. 2010. What is the "New Public Health"? *Public Health Reviews* 32 (1): 25–53.

United Nations (UN). 2015. *About the Sustainable Development Goals*. https://www.un.org/sustainabledevelopment/sustainable-development-goals/. Accessed 8 June 2019.

Walker, Zoi. 2020. Here's why Utahns do or don't wear a mask during the pandemic and what physicians have to say about it. *Salt Lake Tribune* [online] (July 28). https://www.sltrib.com/news/2020/07/28/heres-why-utahns-do-or/. Accessed 1 Aug 2020.

Wertheimer, Alan, and Franklin G. Miller. 2008. Payment for research participation: A coercive offer? *Journal of Medical Ethics* 34: 389–392.

Wiley, Lindsay F. 2012. Rethinking the New Public Health. *Washington and Lee Law Review* 69: 207–272.

———. 2016. The Struggle for the Soul of Public Health. *Journal of Health Politics, Policy, and Law* 41 (6): 1083–1096.

Wiley, Lindsay F., Wendy E. Parmet, and Peter D. Jacobson. 2015. Adventures in Nannydom: Reclaiming Collective Action for the Public's Health. *Journal of Law, Medicine and Ethics* 43: 73–75.

Wilson, James. 2009. Towards a Normative Framework for Public Health Ethics and Policy. *Public Health Ethics* 2 (2): 184–1941.

Wing, Steve. 1988. Social Inequalities in the Decline of Coronary Mortality. *American Journal of Public Health* 78 (11): 1415.

World Health Organization (WHO). 2003. *Controlling the Global Obesity Epidemic* (press release April 23). https://www.who.int/nutrition/topics/obesity/en/. Accessed 31 July 2020.

———. 2019. *Constitution*. https://www.who.int/about/who-we-are/constitution. Accessed 31 July 2020.

———. 2019a. *WHO Called to Return to the Declaration of Alma-Ata*. https://www.who.int/social_determinants/tools/multimedia/alma_ata/en/. Accessed 31 July 2020.

———. 2019b. *Sustainable Development Goals*. https://www.who.int/sdg/en/. Accessed 31 July 2020.

———. 2019c. *Global Health Observatory (GHO) Data*. https://www.who.int/gho/en/. Accessed 31 July 2020.

———. 2019d. *Global Health Observatory (GHO) Data: Noncommunicable diseases (NCD)*. https://www.who.int/gho/ncd/en/. Accessed 31 July 2020.

———. 2019e. *Dementia: Number of People Affected to Triple in Next 30 Years* (News release 7 December 2017). https://www.who.int/en/news-room/detail/07-12-2017-dementia-number-of-people-affected-to-triple-in-next-30-years. Accessed 31 July 2020.

Zimmerman, Emily B., Steven H. Woolf, and Amber Haley. 2015. *Understanding the Relationship Between Education and Health*. Washington, DC: Agency for Healthcare Research and Quality. https://www.ahrq.gov/professionals/education/curriculum-tools/population-health/zimmerman.html. Accessed 31 July 2020.

Chapter 7
Public Health, Communities and Consent

7.1 Introduction

What is or should be the role of consent in public health? Should the role of consent be different if the public health action involves surveillance rather than efforts to change behavior to improve health? Should either individual or community consent be required for the exercise of surveillance? If community consent is required, what constitutes the community and how might it give consent? By definition, public health serves the many, not the one, or at least not only the one. Yet individual choice is the locus of prevailing models of informed consent to health care. So how does consent apply to public health, if at all? And how should it apply? If individual consent should not apply to some or all activities of public health, should there still be consent-like aspects in the process of conducting these activities?

As we discussed in Chapter 6, writers of a libertarian bent such as Jessica Flanigan (2013) answer that public health activities should be analogized to individual health care, and that public health may not interfere with individuals without their informed consent, just as health care providers may not impose care on unwilling patients even when the imposition involves sharing information, unless what is at stake is preventing clear harm to others. At the other end of the spectrum, some insist that individual health care and public health are entirely different enterprises and while consent may be required for the former it is irrelevant to the latter. Still others explore new community-based or population-based understandings of what group consent should and can mean. This chapter explores how the ethical considerations of respect for persons that lie behind informed consent can be reflected in the roles of public health, groups, communities, and populations in surveillance.

Core to the debates about consent are potentially contrasting paradigms of public health, health care research, and individual health care. Ethical frameworks developed in bioethics have successfully argued that in medical research and in the practice of medicine consent should be exercised by individuals, not by groups. By

J. G. Francis, L. P. Francis, *Sustaining Surveillance: The Importance of Information for Public Health*, Public Health Ethics Analysis 6, https://doi.org/10.1007/978-3-030-63928-0_7

189

contrast, public health has the health of communities and populations as its long-standing mission. Whether these more individualistic approaches of health care ethics and the ethics of research should be applied to public health, even when its activities consist of surveillance rather interventions to effect changes in behavior, is the topic of this chapter.

Public health takes as its focus of responsibility the community or the population, not the individual. As described by the Bloomberg School of Public Health at Johns Hopkins University (2019):

Here's a good way to describe the essence of public health.

> In the medical field, clinicians treat diseases and injuries one patient at a time. But in public health, we prevent disease and injury. Public health researchers, practitioners and educators work with *communities* and *populations*. We identify the *causes* or disease and disability, and we implement largescale *solutions*.

Public health is distinct from medical research and health care in its authority to exercise police powers to act for the general welfare. This difference may give rise to both instrumental and non-instrumental reasons for exploring consent. The instrumental assumption is long-standing that public health will enjoy greater success in securing its goals if it does not rely solely on the exercise of the state's authority. Instead, more may be gained for the public health mission by engaging members of the community in listening and in developing shared goals. Some argue that this may be best done through a consent or consent-like process on the part of community leaders or community members. Public health and medical research are in some respects quite similar in their use of data from and about individuals. Like non-interventional medical research with human subjects, much public health surveillance and analysis uses data from and about individuals to develop general knowledge; the difference is that the focus of developing the knowledge may be improving the health of populations rather than the treatment or health care delivered to individuals, although these lines readily blur. These similarities lead some to conclude that consent is required for public health uses of data while others contend that community engagement should be conceptualized differently from individual informed consent.

Beyond what might be needed for the effective exercise of public health authority is its legitimacy. Non-instrumental reasons for exploring consent may begin with assumptions that exercise of political authority over the individual requires the consent of those governed. The need for legitimacy is no less pressing when the activities involve efforts to protect or improve health. However, because their health is also a feature of individual persons, the relationship between consent of the governed and individual consent is complex and intertwining. Additional layers of complexity arise when individuals are members of communities or groups in addition to being subject to political authorities.

This chapter explores four basic themes about consent in public health surveillance. First, community and population differ and have each been given varied meanings in the public health literature. These differences have implications for ethics in the conduct of surveillance. Second, cooperation and volunteerism have been judged to have value in the conduct of public health surveillance, but their role is complex and evolving. Third, the goals of public health are substantively different from the goals of personal health care and this has implications for any requirements of consent. Despite these differences, a common theme of respect for persons underlies both consent to individual health care and the involvement of individuals in public health decisions. Finally, that public health surveillance and interventions cross national borders may have implications for understanding how the mission of public health may be furthered or hindered by surveillance.

7.2 Public Health, Communities, and Populations

Public health practice and expectations have long been associated with community as a focal point (e.g. Goodman et al. 2014; MacQueen et al. 2001). Over the past half century an important theme in public health has been the cultivation of cooperation with communities in conducting surveillance (Public Health Leadership Society 2002). But there is not a great deal of agreement within the public health sphere on a single preferred account of what "community" means. There are also important differences between understandings of public health, community health, and population health (Bresnick 2017). Sorting out these differences is a critical first step to understanding the roles consent might play in surveillance.

7.2.1 Public Health

In the public health literature, one primary use of the term "public health" refers to the activities of organized public health agencies. As we explained at the beginning of this volume, surveillance conducted by these agencies—or others delegated authority by them—is the initial primary subject of this volume. "Public health agencies" in this sense are defined by grants of political authority. There are local health departments, state health departments, and national health departments. Pan-national organizations are devoted to public health such as the Pan American Health Organization or the European Centre for Disease Prevention and Control. The authority of these entities may be defined by treaties or conventions such as the Maastricht Treaty establishing the European Union or the Pan American Sanitary Code and Additional Protocols. At the global level is the World Health Organization, currently with 194 member states. Members of the United Nations may join the WHO by formal notification to the UN secretary-general that they accept the WHO constitution; non-members of the UN are admitted to the WHO by majority vote of the World Health Assembly, the decision-making body of the WHO. These various agencies engage professionally in public health research and practice. The "publics" that they serve are defined by their reach of recognized authority, from the local to the nearly global.

Public health as thus defined jurisdictionally is different from either community health or population health (e.g. Faden and Shebaya 2016). There may be many communities within a single public health authority and there may also be communities that transcend single authorities and require coordination among them. Depending on how "populations" are defined, they, too, may transcend or exist within the jurisdictional scope of public health authorities.

7.2.2 *Communities of Geography and Communities of Interest*

Public health has long sought to engage the communities they serve in participatory programs and research. Community-based participatory research (CBPR) has become a favored methodology especially for research aimed at community health improvement. In this literature, community may be conceptualized as everyone living within a given local jurisdiction (e.g. Stoto et al. 1996). Or, "community" may be understood to refer to groups within and perhaps overlapping a jurisdiction's political boundaries but nonetheless sharing important features that may need to be recognized for successful engagement (e.g. Israel et al. 2013). Recognizing that "community" was given different meanings in CBPR, MacQueen and colleagues sought to understand empirically what participants in this research themselves regarded as characteristics of a community. Based on their findings, MacQueen and colleagues proposed this core definition of community: "*a group of people with diverse characteristics who are linked by social ties, share common perspectives, and engage in joint action in geographical locations or settings.*" (MacQueen et al. 2001, p. 1936). These researchers note, however, that different elements of this definition may prove more salient depending on the context; shared social ties may be far more important in some communities than diverse characteristics, for example. Rajan (2019, p. xiv) similarly defines community in the global economic perspective as unconstrained by size but bound by locality: "a community is a social group of any size whose members reside in a specific locality, share government, and often have a common cultural and historical heritage." Rajan thus explicitly rejects the use of the term "community" to describe virtual communities or national religious communities in favor of the idea of a community as a group whose members live in proximity to one another and share government and heritage.

But there are challenges to this understanding of community as spatially bounded. In an age of global communication and movement, people who share an identity, an interest, or a health condition with people living some distance away may have less in common with their proximate neighbors and more in common with people who share the identity but live hundreds of miles away. Contemporary technology vastly augments and compounds this ability to share identities across distance. Identities such as "gay" or "feminist" or "vegan" may achieve far greater salience than identities such as "New Yorker" or "Parisian" or "southern." Language such as "the black community" or "the disability community" or "the gay community" is common. If these are identities that achieve salience—even if only through the Internet—this might seem to establish their claims to being communities beyond the limits of geographical space. On the other hand, people within a given geographical space may have very little in common, especially if borders are fluid and people frequently move in and out. Illustrating how geography may not establish commonality, MacQueen and colleagues observed that their finding that "diversity" was part of how their research participants understood community may have been influenced by the fact that their subjects were gay men in San Francisco, an international meeting ground for gay men (MacQueen et al. 2001, p. 1935).

In the practice of public health surveillance, a shared identity locally grounded may take precedence over the shared preferences of people at a distance. A critical reason why geographically defined community matters in public health surveillance is the role played by the exercise of powers by local governments. In the U.S., this power to act for the general welfare is termed the "police power" and it belongs to the states or local entities as their delegates, not to the federal government. Public health has long relied on local governmental authorities to require notification of listed communicable diseases and to enable early reporting of new communicable health threats. The potential integration of social and political dimensions of local life only strengthens the appeal of geographically defined community as a basis for public health surveillance. Public health may be able to work through local spatially defined communities to gather information that might otherwise not come to light. Understanding the impact of local cultural or religious practices and preferences for matters such as bodily integrity or diet can play major roles in the ability of public health to encourage immunization or to achieve behavioral changes to address NCDs such as diabetes. Conversely, local attitudes towards behavior judged to depart from local practice, such as sexual practices judged at variance from local values, may suppress information or access to medical care and erect barriers to public health interventions. In these ways, geographically defined community can serve as a building block for or a roadblock to surveillance and other public health activities.

Public health agencies need local communities to respond to their efforts. Ideally communities should be cooperative and attentive in encouraging their members to share personal health data as well as being attentive to reasonable advice given to the community by public health authorities. But ideals of responsiveness may be questioned for their efficacy and their justification. Is there any reason, for example, to assume that communities, however defined, will or should respond in a cooperative fashion to public health? Or, are communities likely to respond in ways that are deeply distrustful and disengaged from collaborating with public health authorities? What if the community in question lacks identity, or if whatever identity is claimed for it arises outside the "community" as an identity imposed or attributed to a set of people but not clearly shared by them? Even shared identities do not mean that the willingness of individuals to be helpful may be shaped by a local leadership (be it political, economic, religious or social) that recognizes and accepts public health competence and authority and enjoys the confidence of the local community. Answering these questions will require working through complex interrelationships between community as defined by local geographical space, community as defined by shared identity, and the reaches of public health authorities.

Moreover, community cohesion may be sustained, strengthened, or weakened by the nature of individuals' relationships to the community. Individuals may find themselves in communities where they have lived since birth. They may be regarded by others as community members. Or, they may enter communities by affirmative acts such as immigration, taking up an occupation or a job, or signing up for a group such as a disease-related social media site. Such acts may be sustained or atrophied by the depth of commitment on the part of community membership more generally. Further affecting community cohesion is that people may be members of multiple communities, especially when communities are defined by identity rather than geographical space. Mobile populations may be members of more than one geographical community, too; this characterization could apply to migratory farm workers, workers in seasonal resorts, or wealthy owners of second homes.

Often important as well to community cohesion are the attitudes of the larger population to the community. Members of the community may be limited in various ways from participation in the life of the larger state or nation as a whole. Exclusion may strengthen community members' own resolve to define themselves against the larger, external population. But the external population may also celebrate the community, which in turn may reinforce its value to the people who make up the community. For public health, what matters is how the cohesion of a community may facilitate or obstruct sharing information or engaging in other activities.

Geographically defined communities may or may not have identifiable leaders other than the leaders of the political jurisdictions in which they are located. If most community members share a common religion, the faith community may have a recognized leadership structure. Elected officials from a cultural or ethnic background that is common in the community may be seen as spokesmen or women for community members sharing that background. Business leaders or others who have achieved success or fame may also be turned to for advice or consultation about the community. However, while these leaders may speak for the views of many in the community, they may not speak for all. Nor do they have the status of recognized political authority or head of an organized group that can claim as members all who live within a spatially defined community.

Three different approaches to community may thus be distinguished for considering the roles of consent:

- community as everyone living within a local political jurisdiction (drawn from Stoto).
- community as one or more groups of people within a geographical area that are linked by social ties, share common perspectives, engage in joint action in these

geographical locations or settings, and that may or may not be diverse in other
ways (drawn from MacQueen)
– community as based in common interests or identity (drawn from references to
possible identity communities such as "the black community" or "the disability
community" or "the Deaf community").

7.2.3 Populations

Populations may be distinguished from communities in any of these senses. A population is a set of people possessing a given characteristic that singles them out for
public health attention. For example, all U.S. citizens, all people living today, or all
people of German ancestry are potential populations. Studies of population health
examine patterns of health outcomes or factors affecting health among the defined
population or population subgroups (e.g. Hertzman and Siddiqi 2009). For purposes
of surveillance, populations are often defined by health risks such as obesity that are
inclusion criteria independent of local administrative or geographical boundaries (e.g. Chunara et al. 2013). Everyone at risk of HIV is a population in this sense,
as is everyone at risk of diabetes from obesity. Populations thus identified by risk
susceptibility may have shared characteristics that give rise to the risks but may not
have much else in common with one other. Collecting data about such populations
may face different challenges than collecting data from a geographically defined or
identity defined community. Whether the focus of surveillance should be population
in contrast to community is determined in part by which takes priority: the disease
of interest; the community of geography, identity, or interest; or some mix of these.

In some ways, in today's world there may be closer relationships between populations, interest- or identity-based communities, and geographically defined communities than in the past. Shared risks may overlap with identities or with
geographies. Features of identity such as race, ethnicity, religion, gender, or sexual
orientation may map onto geographical communities in some cases or transcend
these communities in other cases. Any of these may also map onto or transcend the
jurisdictions of established public health authorities. The conventional expectation
in public health practice is that local governments, in the U.S. often at the county
level, serve as community leaders during public health threats and have ongoing
administrative responsibilities for many aspects of the public health mission. These
political definitions of community often are linked to social, cultural or other values
that prevail within but also transcend local political boundaries.

Public health agencies may seek to delineate communities or populations
depending on what is of interest to their goals of promoting health. However, these
delineations may not fit well with the individuals composing the groups thus delineated. Public health officials may seek to argue that populations or communities
should be defined by a given set of characteristics such as sexual preference,

religious commitment, racial or ethnic identification determined by legislatively determined census categories, age cohort, or generation, but whether they can do so persuasively is another question.

7.3 The Changing Landscape of Groups: Cooperation and Volunteerism

Organized groups may be further distinguished from communities and populations. Groups must be organized in some way, with an identified means for entering and leaving the group but within this characterization, there is a range of possibilities. Groups may be organized around a service, an activity, a health issue, or any other shared concern. Groups may be longstanding, as with many social clubs, fraternities, or religious organizations. Today, groups may be organized to accomplish a short-term goal and facilitated by social media postings. Groups and their members may be present in communities or in populations of interest to public health investigations. The value of groups for public health is that they can be a ready source to facilitate information gathering, communication, and conversations about health issues. Groups may also have leadership structures that be helpful sources for consultation. Groups may, however, also be sources of resistance to public health or of behaviors that may prove troubling, such as resistance to vaccination.

The United States in the nineteenth century and the first decades of the twentieth century was a landscape of towns and villages. Groups of many kinds flourished within this landscape. This multiplicity of organizations was observed by Tocqueville in 1835 (2006 ed.) as the distinctive feature of American society. Small town residents joined in or organized formal organizations such as clubs and fraternal associations (Wuthnow 2013). These organizations combined charitable endeavors, served as forums for community leadership, and acted as advocates for policy change. Notably, some of these groups were founded by women and addressed issues of importance to women such as suffrage. These organizations played extensive roles in the advancement of public health initiatives such as temperance or the campaigns against STIs during the Progressive era described in Chapter 3. Overall, a good number of men and women in the U.S. spent their time when they were not at work or with their families in the membership clubs and organizations found in every moderately-sized town.

In many ways, the voices of American communities were to be found in such service clubs, fraternal organizations such as the Masons, politically minded advocacy organizations such as the Grange, or the trade union movement. Perhaps one of the most successful of all of these was devoted to public health, the Woman's Christian Temperance Union, formed with the political goal of banning consumption of alcohol. The WCTU at its height had four hundred thousand members; today, its membership numbers fewer than twenty thousand. Weekly meetings of many of these organizations were forums where public health authorities could report to communities on health conditions and seek support for public health campaigns.

The respective memberships of such organizations were largely separated by race, ethnicity, gender, and religious affiliation until relatively recently. This darker side of the presence of groups in American society cannot be forgotten. The WCTU did not admit Jews or Catholics for decades. And at its height the racist Ku Klux Klan had as many as five million members and engaged in a wide range of charitable activities (Rothman 2016). The challenge for public health institutions in working with groups is to balance this darker side while recognizing that groups may have useful streams of information. Public health may want to establish ongoing relationships with groups to pursue public health goals, while at the same time the groups have goals that contradict public health values or even goals intended to do harm to other communities and populations locally or nationally.

Political theorists reflecting on this role of groups emphasize the important of non-state actors in both supplementing and balancing the state. For example, Von Gierke (1939) argued that a good society was composed of fellowships such as guilds that played a regulatory role in the economy and in maintaining social well-being. He lamented the growth of state institutional actors at the expense of non-state actors or organizations described as fellowships in reducing the role of voluntary organizations that were independent of the state. The Canadian theorist Will Kymlicka defends the role of groups such as tribes in a liberal political society; his view is that so long as people may leave, groups may depart internally from the more general values of the liberal state with respect to health practices (Kymlicka 1996).

A special case of groups is that of indigenous groups having a level of political authority in addition to their organized structure and claims to cultural rights. Under the International Covenant on Economic, Social and Cultural Rights (to which the U.S. is a signatory but not a state party), states parties recognize the "right of everyone to the enjoyment of the highest attainable standard of physical and mental health." (UN 1967, Art. 12 §1). The 2007 UN Declaration on the Rights of Indigenous Peoples recognizes the right of indigenous people "to be actively involved in developing and determining health … programmes affecting them and, as far as possible, to administer such programs through their own institutions (UN 2007, Art. 23). The Declaration also recognizes rights the highest attainable standard of physical and mental health and to the maintenance of traditional medicines and health practices (UN 2007, Art. 24), as well as to the right to intellectual property over cultural heritage and traditional knowledge (Art. 31, §1). How these provisions apply to surveillance is not specified. In the United States, CDC works with tribes for surveillance of environmental exposures, infectious diseases such as tick-borne Rocky Mountain spotted fever, and even in some cases behavioral risks. These efforts take place in collaboration with tribes. CDC also links with the Indian Health Service to improve the quality of national surveillance data and the extent to which it is representative of tribal populations (CDC 2017, p. 21).

In sum, the contemporary organizational landscape in the U.S. is populated with organizations that in many respects are influential from the local to the global. These organizations vary considerably in structure and relationships to state and quasi-state entities. Voluntary membership organizations have in recent decades contracted in membership and in their attributed leadership role. Some such locally or nationally based organizations that seek a voice in public health-related initiatives do remain, however. For example, the Sierra Club claims a million members and is active in public health anti-pollution campaigns.

Broadly described as non-profits, a million and half voluntary organizations exist in the United States today. Since the 1990s and more so in recent decades, loose confederations of groups have arisen with greater ease to respond to non-communicable as well as communicable threats to health and wellbeing such as natural disasters, gun violence, compromised water supplies, opioid addiction, or coerced sexual relations. These issue-based organizations capture considerable public attention from social media and traditional media alike. Their creation has generated loosely formed groups that can span the county and mobilize large numbers of volunteers to help those in need and to support political intervention to accomplish specific goals. Black Lives Matter and the MeToo movement are prime example of this phenomenon. Such formation of short broad coalitions built around the power of an immediate shared concern is enabled by social media. The recognition that one quarter of the American population will volunteer for a wide range of services to others (Bureau of Labor Statistics 2016) has reduced traditional constraints to action.

That we live in age of non-state actors is particularly evident in the expanding world of public health (WHO Director General 2019). In the arena of public health, many organizations, some more organized and some less loosely organized, are of value in disseminating information about public health-related challenges. Perhaps in the expanding remit of public health, Greg Consalves is a 2018 MacArthur "Genius" awardee who has developed research tools in academia to draw creative and unusual connections between research and social interventions. His efforts have long been directed at establishing productive relationships between public health research and activists to address ongoing rates of HIV infection, links between violence against women and lack of access to indoor sanitation, and the failure of mining companies and the South African government to compensate families for occupationally-related disease.

Following Robert Putnam's image in *Bowling Alone* (2000), a lively literature has called attention to the declining membership of organized clubs and societies. What appears to have happened is the rise of nonprofits that blur relationships to the state, the community and to the force of contemporary social media. In short, Von Gierke might have been more prescient to map out the complex relationships among non-profits, the state, and volunteerism that shape to an impressive extent public health initiatives today.

For community involvement in sustaining the public health mission, the identification of leadership is complex but critical. Using leadership of local organizations is one strategy of consulting and securing support for public health surveillance. Local leadership communities may not coincide with local governments and, even when they do coincide, may not reflect the preferences of the local population. If the goal for public health is a more nuanced account of the contours of opinion, then sorting out leaders who capture the range of concerns in the population is core to likely success. If the need to undertake a public health program is so urgent that consultation cannot be obtained in the time frame available, as with rapid spread of infection in a community where vaccination rates have plummeted far below herd immunity, community leaders can be consulted as a substitute. As we just discussed, however, it is neither clear who these leaders might be nor clear as to the authenticity of their ability in the addressing the health concerns of the identified population. This does suggest that consultation is important but with whom and under what conditions is from time to time elusive.

7.4   Consent: Public Health and Individuals

Any discussion of consent and public health must start with the recognition that public health is not the same as individual health care. This chapter began with the observation from the Bloomberg School that clinicians treat one patient at a time while public health prevents disease or improves health in populations. Nevertheless, the public and the individual cannot be fully separated, just as the public sphere and the private sphere abut, overlap, and intertwine. Addressing an epidemic threatening the public may require treating individuals who are the sources or likely locations of spread. Case identification and contact tracing has remained essential in the fight against COVID-19. At the same time, community factors clearly matter, too, such as safe workplaces. Protecting the health of the individual may be impossible or impracticable without addressing the health of those in the surrounding area or community. Nor can consent to personal health care be entirely teased apart from consent in the context of public health. Public health authorities stress the importance of individual consent in building trust among the population served while at the same time stressing the limits of consent in addressing needs for surveillance and protection.

Also in the background of the discussion of consent and public health is the unlikely prospect of insulating individuals who might decline to be surveilled or to accept the terms of a public health community-based response to prevent disease spread. Any view of the role of consent must take into account this observation that individuals cannot be isolated as unique health islands but must be in some way subject to the processes of public health. Individuals are interconnected most obviously for contagion. But as we saw in Chapter 6 about the new public health, they may be interconnected at least for information about other dimensions of health as well. That no one is a health island poses an immediate challenge for insistence on individual consent to public health. Insistence on unanimous individual consent has the potential to erect insurmountable barriers to any successful public health effort.

Political scientists refer to this ability of an individual to block effective collective action as the "Polish veto" or "liberum veto" after a practice in the Polish-Lithuanian Commonwealth that enabled debate to be suspended and legislation nullified on the request of any envoy to Parliament. In eighteenth century Polish-Lithuania, the veto was instituted to safeguard the liberties of nobles against an absolutist monarchy and maintained as the "apple of liberty's eye" (Lukowski 2012,

p. 71). The risk of legislative paralysis was avoided as long as there was effective consensus about restraints on use of the veto. The solution to eventual legislative breakdown, according to Polish historian Jerzy Lukowski, was to address the mistrust that had arisen between the monarch and the aristocracy by enacting constitutional reforms that enshrined protections for liberty. In the end, Lukowski writes, "Unanimity supposedly continued to hold sway, no matter how improbable this must have been in assemblies numbering dozens or even hundreds of participants. Unanimity remained the ideal actively pursued by *szlachta* [noble] society: to attain it, safeguards were more important than efficiency or effectiveness" (2012, p. 96–97).

Such political experience with the individual veto is instructive for understanding the role of individual consent to public health activities. On the one hand, making individual consent necessary for any public health activity risks great public harm. Even if consensus were possible to negotiate, it takes time; in contexts of public health emergencies, timely information and action may be essential. Mistrust is endemic when people are afraid of the spread of deadly disease from one another or from sources unknown. Addressing mistrust is at the heart of sustaining surveillance in these circumstances, just as it was in the Polish legislature of the eighteenth century. Consent, or consent-like strategies, may be part of solutions to resistance to public health efforts.

7.5   Individual Informed Consent: Models from Bioethics

This is an age of ubiquitous consent on both sides of the Atlantic. Insistence on consent is ever-present in discussions of private affairs from sexual relations to marriage and the creation of families. The ease with which consent can be given and the consequences that may result remain contested, however. Notice and consent substitute for regulation to protect individuals from risky products or activities, from sharing information over the Internet to hang gliding. Consent also creates contracts and absolves manufacturers from liability. Consent may occur almost without people realizing it, with a click of an "I agree" button to share content over the Internet or an eager unwrapping of a shrinkwrapped package without reading the notices it contains. Consent may occur apparently quite readily even when important rights are involved, as when an employee signs a contract of employment without reading that she has ceded her right to go to court rather than be subject to arbitration if a dispute arises about her job. In some jurisdictions the reaction has been to significantly tighten some consent requirements, as in the European Union's Data Protection Regulation or the development of consensual relationship policies on US college campuses. Autonomy has grown steadily within and outside the family and in other autonomous or partially autonomous groups consent has become the common expectation of the day.

Chapter 3 described the development of individual informed consent in bioethics and its impact on testing for HIV. This autonomy-based paradigm locates the process of consent in separate individuals; assumes they are aware of their values, conditions, and options; and arms them with the knowledge, time, and skills to assess which course of action will best serve their values. Through support for confidentiality and control of information, the paradigm of individual consent is extended to information as well as it was for testing for HIV.

Informed consent is not a panacea for protecting individuals, however. The paradigm admittedly idealizes, in well-recognized but problematic ways (e.g. Faden and Beauchamp 1986; Kim 2019). Many decisions about personal health care are made within asymmetries of knowledge, resources, power, and dependency. Patients may be in pain, in fear of suffering, or faced with death. Decisions may need to be made quickly without time for calm or careful reflection. Patients without direct experience may find it very difficult to imagine what life would be like in an altered condition; people notoriously express different views about what they believe would be their quality of life with hypothesized disabilities than they do about their lives with actual experience of the disability.

Recognition of individual embeddedness in relationships and communities presents resources for addressing the inadequacies of informed consent in enabling individuals to seek to realize their values in receiving personal health care. Patients' family or friends may play mitigating roles in reducing the flaws that plague effective consent. Relational theorists have pointed out that individuals exist in interrelationships with others that are constitutive of their identities and theories of their good. Individuals also may not make decisions fully on their own but with support and participation from others.

In addition to families and friends, support groups, either real or virtual, provide forums for sharing information and experiences, hopes and sorrows, and strategies and resources. The popularity of Facebook groups for parents of children with birth injuries or PatientsLikeMe for individuals with comparatively rare conditions illustrate how decisions about health may become embedded in interpersonal contexts. It should not pass without notice that support groups or Facebook groups may be groups or possibly even communities of at least some of the types delineated above. Groups and communities potentially offer resources to address at least some of the asymmetries of individual informed consent. The possibility also should at least be open to discussion that public health may have resources, including trusted experts or leaders, to counter asymmetries in individual consent. Whether groups, communities, or even state public health agencies function supportively or repressively, and give information or disinformation, is of course open to question. The only point at this juncture of the argument is that informed consent has insufficiencies that resources beyond the individual may be helpful in addressing.

Writers about informed consent have increasingly sounded the theme that it should be regarded as a process rather than an event. "Shared" decision-making under which physicians and patients develop mutual understandings and knowledge and work together to achieve care plans that reflect the patient's values is a common description of this process. Here, too, there may be insights for individual consent from public health. Decision-making processes involve not only working together but also adjustment and compromise, trial and error, and revision and re-revision.

Sometimes, what matters more is the acceptance, thoroughness, and fairness of the process—not the result. Seen in this way, the process of individual informed consent is not the excavation of a "right" choice for a given patient. Rather, it is achievement of a plan that has come about in a good way. The decision may not be final, either, but subject to evaluation, re-evaluation, revision, and re-revision.

The informational function of informed consent is relevant here, too. *Informed consent is about enabling people to have the knowledge they need to make the decisions they face.* It thus requires judgments about what information is needed and how that information can be shared to improve understanding. Judgments about information and how it may be shared are themselves complex and contested; as Chapter 2 described, even scientific judgments are not univocal or settled. The process of subjecting recommendations to transparency, examination to avoid conflicts of interest, and assessment and reassessment is not simply individual. Of course, decisions about relevant information may be politically motivated and problematic; state legislatures in the U.S. have enacted legislation notoriously requiring physicians to read discredited information to women seeking abortions, such as that abortion causes depression or increases risks of cancer. Once again, the point here is not that public intervention is always or even usually benign, but that it may have insights and resources to contribute to the process of individual consent.

In individual health care, one asymmetry remains on the patient's side, at least for the most part in many societies: the right to say "no." To be sure, as a practical matter this right may be differently achieved, honored in the breach, or ignored for patients with different vulnerabilities and resources. But common law jurisdictions recognize imposing health care on people without any consent as the tort of battery (*Schloendorff* 1914), at least absent special justifications such as emergencies or imprisonment. Vaccination, too, has been long recognized as an exception in many jurisdictions, although the requirement remains controversial. In *Jacobson v. Massachusetts* (1905), the U.S. Supreme Court opined over a century ago that "The police power of a State must be held to embrace, at least, such reasonable regulations established directly by legislative enactment as will protect the public health and the public safety (p. 25)." Moreover, the Court reasoned,

> ...surely it was appropriate for the legislature to refer that question [of what should be done in a health emergency], in the first instance, to a Board of Health, composed of persons residing in the locality affected and appointed, presumably, because of their fitness to determine such questions. To invest such a body with authority over such matters was not an unusual nor an unreasonable or arbitrary requirement. Upon the principle of self-defense, of paramount necessity, a community has the right to protect itself against an epidemic of disease which threatens the safety of its members (p. 27).

That this decision was later infamously applied to permit eugenic sterilization of those believed to be intellectually disabled (*Buck* 1927) has raised continuing questions about the scope of such public health authority but not about its underlying constitutional authority to act for the overall public welfare.

Libertarian theorists, however, urge that the asymmetric ability to say "no" should remain on the part of members of the public for to the extent possible in public health. Except in the clearest cases of self-defense on the part of the public, these theorists argue, people should be able to opt out of public health activities. This position has been urged especially for interventions in support of the concerns of the new public health, as discussed in Chapter 6. But it has also been extended to gathering information needed for surveillance, at least to the requirement that people should be permitted to "opt out" of any requirements to share information about themselves.

Jessica Flanigan (2013, 2017), for example, argues that exactly the same commitments that justify anti-paternalism in bioethics extend to anti-paternalism in public health and to regulation of pharmaceuticals. People who are entitled to make unwise judgments as patients, she says, should surely also be entitled to make these same unwise judgments as consumers. After all, "consumers and patients are the same people" (2013, p. 173). "Because it is wrong for a physician to substitute his judgment for a single patient's, it is even more wrong for a policy maker to substitute his judgment for an entire population's" (2013, p. 175). Flanigan does limit her argument to coercive interferences; she supports the view that governments may act to ensure that people have access to adequate information and may also engage in nudging to counteract known cognitive biases. Otherwise, she contends, people may not be made to act for their own good, even when that good involves sharing information about themselves. That the authority intervening is democratically legitimated does not solve the underlying justificatory problem for Flanigan: if physicians may not impose health care on patients or reveal their health records without their consent, that the legislature has authorized them to do so should not make a difference. Both public and private power, she contends, must answer to the underlying rights of individuals, rights which take precedence unless the balance of moral reasons counts in favor of coercion.

As pointed out in Chapter 6, however, information poses a problem for this analysis, if information about some is needed for others. There are strong non-paternalistic reasons for surveillance. Defenders of individual control over information reply that depending on the circumstances some may be able to opt out of sharing "their" information without compromising the information available to others. According to this analysis, information is analogous to herd immunity: society does not need universal vaccination in order to be protected against the spread of infection, nor does it need universal information to protect the public against dangers to health. To be sure, both vaccination and information requirements to protect the public vary with the circumstances. Depending on the contagiousness of a disease, the percentage of the population needing to be vaccinated to create herd immunity will vary; for measles, 90–95% of the population must be vaccinated, but for polio the number is lower, for example (e.g. Silverman 2019). Similarly, depending on what is needed for data to be sufficiently representative of the population, the percentages that can opt out without compromising the analysis will vary. The distribution of opting out percentages matters, too; if a high proportion of people from a particular group decline to share, the data will be compromised for that group, just as herd immunity may not be created in a closely-knit population subgroup. Moreover, just as there are situations in which particular individuals must be vaccinated—for example, family members of an index case or travelers leaving an infected area—so there may be situations in which information about particular individuals is necessary, for example to determine the significance of a genetic variant in a relative or to trace the transmission of an infection.

Thus, opt out strategies may not always be available without compromising public health. Rigid insistence on individual informed consent as a necessary condition for public health surveillance is not an acceptable position, as Onora O'Neill has argued (Manson and O'Neill 2007; O'Neill 2003) along with many other scholars (e.g. Berg 2012). Nonetheless, important insights are to be gained from the idea of informed consent and the underlying values of respect that it was designed to capture. People should be informed. They should be involved in decisions not merely as objects but as participants in the process. They are not—as Kantian autonomy captures—mere means only; they are ends deserving of respect. These points are not merely instrumental, although public health is likely to be met with resistance when people become suspicious that information is being withheld or that they are being used for the benefit of others. Instead, they are core ethical values that reflect why resistance may not only occur but also be warranted. They are values to be realized in the context of non-paternalistic justifications for public health surveillance.

The discussion that follows takes up how these elements of respect—information, participation, and subjectivity—can be reflected in surveillance decisions made by governmental units or decisions involving groups, communities, or populations. These processes are consent-like in some ways, but they are not the individualistic informed consent paradigm of bioethics. Nor are they the adoption of an opt in or an opt out framework for individual participation. The expectation is broadly accepted that consent matters in health care particularly when personal data are shared, and that data sharing is more likely to be successful if the population served believes it can trust public health authorities. The challenge for public health is how to acknowledge the concerns of a community without accepting a commitment to seek individual consent or agreement to gather and share personal data. Recognition is critical of some kind of community responsibility for the data collected and the consequences of its distribution for the individuals concerned (Lee et al. 2012).

In what follows, we develop how each of the different ideas of community sketched above may—or may not—instantiate elements of respect through sharing information, encouraging participation, and recognizing the subjectivity of members as people with their own values. The different ideas of community considered are:

– the political legitimacy of established public health authority
– group membership
– geographically defined community
– interest or identity-defined community
– population as defined by risk or some other characterization of interest.

7.6   Public Health Authorities: Democratic Practice, Political Participation, and the "Consent of the Governed"

Public health functions as an agency of government. One possible strategy for respecting individuals in surveillance is reliance on existing political institutions. Democratic forms of government are generally viewed to present the closest analogy to individual consent, as reflecting "the consent of the governed," so we consider them as the best case here.

The "consent of the governed" has over the past two and half centuries been widely regarded as necessary to establishing a democracy and sustaining its existence. Yet how the requirement for consent of the governed is to be met has never been entirely clear. What consent means in institutional practice has been even less clear. As is regularly pointed out, consent as the right to vote did not extend in any obvious way to the great majority of people living in the British colonies at the time of the Declaration of Independence. Consent as the right to vote was gradually extended to greater numbers of people with the elimination of property qualifications and conditions of servitude, the grant of women's suffrage, and reduction of the voting age to 18. But consent as the right to vote is not universal; depending on the jurisdiction, it may not extend to those who are underage, who have been convicted of certain crimes, who have been declared mentally incapacitated, or who are not residents or citizens.

The relationship between the right to vote and consent is a different matter, however. On one end of the spectrum is the view that voting in an election suffices for actual consent. The act of voting serves as an endorsement of the process and so arguably of consent to its outcomes. Conversely, someone who does not vote might be presumed not to endorse the process but to prefer a different system and hence not to consent. If only a few people vote, the more general conclusion might be drawn that people have taken the initial steps to seek new institutional arrangements for how they are to be governed. Equating voting with consent and not voting with not consenting, however, would be premature. The failure to vote also could be construed as a measure of satisfaction that the existing system is working well and so there is no reason to go to the polls. Or, it might reflect ignorance; political philosopher Jason Brennan (2016) questions allowing irrational or ignorant voters to have any kind of say in political decisions that affect others. Similarly, voting could be construed as an expression of the need for change. The meaning of voting is ambiguous, thus putting into question the relationship between voting and consent.

On the other side of the spectrum is the libertarian view that voting can never replace requirements for individual consent. Jessica Flanigan, whose opposition to the new public health was discussed in the preceding chapter, argues further that the fact that a democratically elected government institutes a coercive policy does not change whether the policy is justified in the first place (Flanigan 2013). If consent is required for someone to undergo a medical procedure, as she claims that it is, democratic authorization for patients to undergo the procedure without their consent does not constitute authorizing them to undergo the procedure without consent. On

Flanigan's libertarian view, it is only permissible for a democracy to coerce when it would be permissible for an individual to do so, and it is only permissible for individuals to coerce when it would prevent them from being harmed by others (2013, p. 176). That the individual has accepted the benefits of living together in society or has in some way agreed to the decision process does not change this underlying limit on when the use of coercion can be justified. Instead, Flanigan thinks, the individual must have consented to the benefits or the process with the coercion as a condition, in order to be considered to have consented to the coercion too.

Libertarian positions about democratic legitimacy such as Flanigan's may be criticized on many grounds. Most importantly for public health, such libertarianism holds a very narrow view about the extent to which individuals' actions affect only themselves and not others. On such libertarian views, someone's diabetes affects only herself whereas someone's measles might affect others. If, to the contrary, we think that someone's diabetes does affect others—or take the more far-reaching structural position that someone's diabetes and whatever effects it might have on others are embedded in a common social framework—then we have grounds for rejecting the line between self and others that underlies the insistence that individual consent is necessary to legitimate any coercive public health measures that extend beyond preventing contagion.

Rejecting libertarianism does not, of course, mean accepting the view at the other end of the spectrum that voting constitutes consent. A more moderate view might be that, in a democracy, legislative oversight reflects the concerns about respect for persons that motivate consent requirements. If in the jurisdiction in question the legislature assigns responsibility for decisions to administrative agencies, then the question would be whether a structure of legislative assignments regarding surveillance to public health sufficiently respects individuals who are involved in the surveillance. In practice, a good deal of surveillance has been conducted at the local government level and local governments perhaps present the most compelling case for effective democratic oversight of surveillance. Oversight allows accountability of elected leaders to their constituents rather than agreement in real time of every constituent to every decision. Nonetheless, even with this level of oversight there are gaps in achieving respect.

One limitation of democratic oversight is that it yields at best a patchwork quilt of yeas and nays to majority decisions accepting the terms of surveillance. People who are persistently in the minority may mistrust and reject participation in surveillance; resistance may be especially significant if these are people who are in the minority on many issues or are from disadvantaged groups. On the other hand, adequate surveillance may be exactly what is needed to reveal disadvantage, as the publication of the disparate impact of COVID-19 has done. Timing is another difficulty, as decisions may need to be made quickly about issues that may have been difficult to predict in advance such as a novel pandemic; consultation risks obstructing these decisions. These limitations might be addressed by regarding public health surveillance as a collective commitment that is ratified by policy makers on behalf of the governed to be performed by agents of the public health department. That is, authorization occurs beforehand at the policy level. Afterwards, failures to object take the place of consultation at the moment.

Whether authorization and apparent later acquiescence are meaningful for people in the jurisdiction is questionable, however. Problems of minority exclusion may remain. In addition, members of the political community at one point in time may be quite different than members at a later point in time. Members of a later political community may not even be aware of—much less have been consulted about—earlier decisions about surveillance policies. Another problem is that some or many members of the political community may be poorly informed at any point in time; unless efforts to communicate are assiduous, surveillance may come as a surprise to some.

Two decisions by the New York City health authorities—one about limiting the size of cups of soda and the other about diabetes surveillance—illustrate how these gaps may emerge and become significant even when decisions can be made without time pressures for immediate action. In 2013, as discussed in Chapter 6, New York city health authorities decided to implement a cap on the size of sweetened beverages. Because neither the city council nor the state legislature had been able to agree on an approach to reducing consumption of sugary beverages, and city public health officials considered these drinks to be a significant public health threat, the Board of Health adopted a regulation prohibiting specified establishments from selling drinks in cups larger than 16 ounces. A successful court challenge determined that the Board's action exceeded its regulatory authority—that is, that the Board had acted on its own rather than under the umbrella of the required legislative action.[1] The court challenge intervened to block what was widely regarded as action by the Board without appropriate democratic oversight.

The decisions about diabetes surveillance, however, illustrate insufficiencies in legislative authorization. Several years before the soda size regulation controversy, the Board had taken a set of highly controversial surveillance actions regarding tests for a measure of diabetes control. It created a hemoglobin A1c registry and required reporting of test results to the registry. It also implemented a pilot study in the South Bronx, an area with particularly high diabetes rates. Under the pilot as originally proposed, patients would be informed of their test results and significance and physicians would be given quarterly summaries of their patients' levels of diabetes control (Chamany et al. 2009). (After opposition surfaced, the pilot was designed to allow patients to opt out, although critics claimed that the opt out process was too difficult for many to use.) These actions were met by extensive ethical criticism for violating privacy, imposing an intrusive form of paternalism, interfering with physician-patient relationships, and discriminating against largely minority residents of the South Bronx. Notably however, while critics cited disanalogies between tracking diabetes as an "epidemic" and listing contagious diseases for surveillance (e.g. Barnes et al. 2007), whether the action was within the legislative grant of authority to public health was not challenged. Instead, the controversies about the registry proposal were all about the conflicts between privacy and public health presented by the registry. Privacy advocate Janlori Goldman (2008) and her

---

[1] New York Statewide Coalition of Hispanic Chambers of Commerce v. New York City Department of Health and Mental Hygiene, 16 N.E. 3d 538 (N.Y. 2014).

coauthors argued that the registry lacked adequate privacy safeguards, threatened the integrity of the physician-patient relationship and would backfire as patients avoided seeking care. Law professor Harold Krent et al. (2008), in a thorough legal analysis of the registry, pointed out its many legal risks to patients but did not ask whether the registry overstepped the authority of the public health agency.

Public comment had been solicited before establishment of the program (Chamany et al. 2009), so there was at least some transparency and opportunity for the public to be heard. However, Fairchild and colleagues describe the process of public consultation as limited and not at all systematic (Fairchild et al. 2007a, b; Fairchild 2006). Major city hospitals and clinicians were consulted, but local and state medical associations were not. The American Diabetes Association was consulted and agreed to support the registry only if patients gave informed consent to participation. During the development of the registry proposal, privacy advocates were not involved, according to Fairchild (2006). Citizen advocates pressed privacy concerns at a hearing, but few public comments were submitted. There is no report of efforts to engage local groups in the South Bronx despite the area's history of community organizing; one study several years later of the South Bronx after the recession reports that the area had significant "bridging social capital among …different cultural and historical worlds" (Parés et al. 2017). Pushback against the registry was the result and subsequent efforts to extend the registry model to HIV were met with stiff opposition. Even though the registry did not draw the challenge of overstepping public health's administrative authority that was later wielded successfully against the soda size regulation, it can be viewed as a failure of democratic oversight. Respect for those governed by surveillance requires more than a general legislative grant of authority and public notice if it is to be sustainable.

7.7 Involving Groups and Communities

Earlier in this chapter, we identified groups, communities bound by geography, and communities of identity or interest as potentially distinct sources for community engagement. Here, we consider whether—and how—each of these might be involved in surveillance in ways that help to generate the forms of respect that underlie claims for informed consent: information, participation, and recognition of subjectivity.

7.7.1 Groups

Groups have the advantages for consultation of recognized leadership and membership structures. Some groups may also have governmental authority—Indian tribes are the best example in the United States. With the advent of social media, groups have become far more fluid, although some with quite extensive structures remain.

Groups may have established methods for communicating with members that enable ready dissemination of information. Groups may also have structures that encourage members to voice their concerns by coming to meetings, commenting through social media, or even talking with other group members. If groups have identified values, religious or otherwise, people holding group leadership positions may be able to give effective voice to these values. This is not the same as "speaking for" the group, which would suggest that the group leader is acting in a representative capacity. Rather, it is giving voice to values that are likely to be shared among group members. Communication here may also be more than one way, if group leaders and group members discuss shared concerns and how values might be articulated in light of these concerns.

An example of involving groups in these ways is provided by recent vaccination controversies in Orthodox Jewish communities in New York. These communities have high rates of vaccine refusal. When they were beset by a measles epidemic in 2019, the response of the City health department was to mandate vaccination within four of the most affected ZIP codes. The result was controversy and resistance. Cantor (2019) urges instead that there should be support for efforts to address the vaccination refusals through groups respected by the community. The Orthodox Jewish Nurses Association (OJNA) is a group founded quite recently—in 2008—to provide a network of support for Orthodox nurses. Membership requires dues; the organization maintains a website, publishes a journal, has local chapters sponsoring social and educational events, and maintains an active social network presence on Facebook (OJNA 2019). In response to the misinformation campaign that had encouraged vaccine refusals in the New York Orthodox communities, OJNA published a booklet explaining vaccine science and OJNA members met with small groups in the Orthodox community, especially mothers who had received antivaccination appeals (Cantor 2019). Community religious leaders also spoke out in favor of vaccination (LaMotte 2019). More so than top-down public health requirements, these efforts have the potential to engage group members in addressing their concerns and countering other communication strategies that are also directed at group members.

7.7.2 Geographically Defined Communities

The vaccination controversy is unusual in the extent to which it affected a particularly tight-knit group. Surveillance, as well as other public health activities, may also involve geographically defined communities. As described above, these communities may share some or all of cultural or religious ties, histories, perspectives, or interactions. They are geographically bounded but not necessarily coincident with political jurisdictions.

The movement for community engaged participatory research (CBPR) has generated a considerable literature about involving geographical communities in research (e.g. Coughlin et al. 2017; Dolgon et al. 2017). CBPR is research, and thus emphasizes consent by both individuals and on behalf of the community, but the literature contains a wealth of materials about engagement. Establishing community partnerships is core to the methodology of CBPR. According to one standard text about the methodology as it is used in health research, nine principles guide CBPR partnerships:

- communities are to be acknowledged as units of identity
- research must build on community strengths and resources to identify concerns
- partnerships must share power equitably in a manner that recognizes existing social inequalities, for decisions at all stages of the research process
- processes must be reciprocal and involve learning on all sides
- there must be a balance between knowledge generation and intervention for the mutual benefit of all community partners
- focus must be on the local relevance of health problems and emphasize an eco- logical approach
- systems development must be addressed through a cyclical and iterative process
- results must be disseminated to all partners and partners must be involved in the wider dissemination of any research results
- the process must be long term and involve commitments to sustainability (Israel et al. eds. 2013).

Several themes in these principles are particularly relevant. CBPR is to be rele- vant to community concerns; similarly, surveillance might attend to issues of par- ticular salience in the community. Power and learning should be shared and reciprocal; surveillance decisions should not be top down only. Benefits should be shared, too. Ongoing communication is critical, especially about what is being learned and how that information is being shared. And surveillance is an evolving and ongoing process.

These principles describe what partnerships with communities should generally be like but do not themselves describe how communities should be approached to achieve these goals or how particular partners should be identified. Addressing these issues, the text suggests that partners may be identified in many ways, but a common starting point must be how residents of the geographical area define their identities. To get to know communities, researchers are urged to contact all the orga- nizations that might give voice to people within the community, from parent-teacher associations and schools, to community safety or housing groups, to environmental justice coalitions (Israel et al. eds. 2013, p. 48). They are also urged to identify com- munity "movers and shakers" through these groups (p. 54). A recognized ongoing challenge for CBPR, however, is to reach beyond service professionals and other policymakers to other community members who may lack time or resources to attend meetings. Providing meals and childcare may help but barriers such as work schedules may still impede community participation. Another challenge is to develop structures for ongoing collaboration, all along guided by overarching principles such as equity and reciprocity. These challenges are also faced by public health as it seeks to involve its community in surveillance.

Fundamental to CBPR is the idea that the community itself has interests and concerns beyond the coinciding interests of individuals making up the community. Whether and how to identify these community-level factors is not easy, however. A literature review of discussions of ethics in CBPR identified community considerations as community self-determination, protection of community values, respect for culture, dissemination of results to the community, equity within the community, and consideration of community needs over individual liberty (Mikesell et al. 2013). In a 2011 review, Shore and colleagues assessed how community concerns are addressed in the process of reviewing the ethics of CBPR research by institutional review boards and concluded that community-level considerations typically are not included in the reviews (Shore et al. 2011). This study surveyed community groups involved in CBPR partnerships involving research with human subjects and found that just over a third did not have processes for deciding whether the community should participate in research. They also concluded that many communities most affected by inequities did not have the resources to create effective review processes. A noted—and notable—limitation of the research was that it did not ask how survey participants themselves defined "community" (Shore et al. 2011, p. S363).

A 2017 report of the Committee on Community-Based Solutions to Promote Health Equity in the United States of the National Academy of Medicine attempted to tackle the problem of involving communities in addressing health inequities. The definition of "community" used for this report was sufficiently broad to include both spatially defined communities and communities of interest: "Any configuration of individuals, families, and groups whose values, characteristics, interests, geography, and/or social relations unite them in some way." (Weinstein et al. eds 2017, p. xxiii). The report describes how [c]ommunity assets can be built, leveraged, and modified and can create a context in which to foster health equity" (p. 9) by developing concrete examples of community-based solutions to health inequities. In selected case studies, the report describes these strategies for engaging communities: getting recommendations for stakeholders and partners from the mayor and city council, consulting local health care centers, consulting neighborhood community organizing centers, consulting local charitable organizations such as United Way or a nonprofit Children's Bureau, engaging local schools, linking faith communities such as Protestant churches and the local Catholic diocese, working with an organization devoted to food security, working with an organization pursuing environmental justice, and contacting local social service agencies, among others. This variety suggests that successful methods for engaging communities will vary with the individual circumstances of communities. Challenges include communication, the time and energy needed for participation in longer-term projects, and changes in political administrations (Weinstein et al. eds. 2017, p. 315). Essential to addressing these challenges are shared visions, hope for a better quality of life, trust, and community agency, according to the report. Without sufficiently charismatic leaders, community engagement often failed. Community partnerships also required "very specific governing practices and structures that were tailored to the needs and makeup of the community being engaged" (p. 319). This analysis suggests that the more structure and leadership exists within the community, the better may be the prospects for sustained engagement of the community in surveillance that is acceptable to them.

7.7.3 Communities of Interest, Communities of Identity, and Populations

Communities of interests or identity may lack geographical cohesion or any recognized leadership or structure. A group bound by an interest—in protesting genetically modified foods, achieving a cure for a devastating disease, or accessing expensive drugs for their condition—may be bound by little else than the interest that brought them together. No community values or ties may extend beyond the interest that is mutually shared. Individuals may be willing to share data to address that interest, especially if the interest is highly salient to them, but addressing other forms of surveillance with communities of interest may be more difficult. Once data use goes beyond the interest that brought these individuals together, cohesion may be difficult to sustain. Without any structure other than the shared interest, it will be more difficult to establish inclusive political channels sufficient to sustain information exchange or participatory engagement.

On the other hand, some communities of interest may have established communication networks, especially through the internet. Facebook pages or other social networking sites may be useful for sharing information about surveillance and its results. Email lists and lists of followers on sites such as Twitter can serve as a way for getting out information about surveillance and its results. Social media can also be a means for inviting participation. Polling, qualitative research strategies, or online focus groups may be possibilities for gauging attitudes of those involved in communities of interest and attempting to respond to them.

Communities of identity such as "the Black community" or "the disability community" may present further difficulties. Communities singled out by an identity characteristic may not be linked in any other way. There may be no established channels of communication to transmit information about surveillance, its results, or other important health information. Public health may rely on media that it believes are more likely to appeal to people who share the interests in question. For example, the Centers for Disease Control and Prevention tried to educate Blacks about HIV/ AIDS testing by advertising on radio stations believed to have high rates of listeners who are Black (Hall et al. 2010). This method of communication is imperfect at best, however. There may be no structures other than existing political institutions to encourage participation. A further challenge is that an identity characteristic used to delineate a supposed community of interest may—but may not—be accepted by all those who are considered as within the community (Jewkes and Murcott 1996). Imposition of an identity characteristic may be met with surprise and displeasure by those who believe they have been mischaracterized or who do not wish to be associated with a community of identity. Communications aimed at people that presume membership in an identity community may backfire as a result. On the other hand, relying on people to self-identify as members of an identity community—as does the census for categories such as black/African American—may be both under- and over-inclusive, especially if people have multiple or mixed identities.

Whether individuals are singled out as members of a community of identity or as a population subgroup may also be unclear. Populations are defined by questions of interest for public health, such as risks. These may or may not coincide with communities of identity such as Blacks in the U.S., residents in the US, or everyone living in areas of high rates of a particular disease such as HIV/AIDS. If a public health official states the importance of working with the Black community on a public health issue such as HIV, that official may mean all Americans who indicate they are Black on a census form, or a set of Black individuals living in a specific geographic space, or perhaps more ambitiously the Black population (however identified) of the nation as a whole. The official also may think of a nationally dispersed set of localities, each one a community that in turn has much in common with other communities. In this example, individuals may share an identity, they may or may not have chosen the identity, and others may have designated the identity for them externally.

Efforts by the CDC to address rates of HIV in the U.S. illustrate how communities of interest, communities of identity, and populations may become blurred. Rates of HIV and rates of new diagnoses of HIV among people who self-identify as coming within the census category black/African American are significantly higher than rates among other racial or ethnic groups (CDC 2019). In reporting this data, the CDC refers to black/African Americans, whites, and Hispanic/Latinx as subpopulations. It also refers to the challenge that "some African American communities have higher rates of some sexually transmitted diseases (STDs) than other racial/ethnic communities," using the language of community rather than subpopulation but here as referring to what might be geographically located communities of Blacks rather than communities of identity. The federal government website HIV.gov describes its informational sheets concerning HIV rates among blacks/African Americans nationwide as "HIV's Impact in the African American Community" (HIV.gov 2019)—here, using the language of a community of interest to refer to the population subgroup of black/African Americans rather than to communities with shared ties. These differences matter. Given the growing knowledge about HIV transmission within groups that share social interactions, blurring the lines between communities defined through geographical linkages, communities of identity, and subpopulations may prove troubling in developing strategies for addressing disease spread (Morgan et al. 2018; Sullivan et al. 2018). But it is also troubling about building trust in surveillance, as strategies for communicating with groups and local communities may be quite different from strategies for addressing subpopulations.

7.8   Movement: Surveillance Crossing International Boundaries

A still further problem for communicating about and engaging people in surveillance is that people move. When they move, disease travels with them. This has been true at least since the early great empires. The Roman Empire in many respects was a success. Once the Romans conquered a kingdom, the new province gained access to trading markets from England to the Nile and as far as India. Roman engineers built new water systems that supplied public baths and running water, put on public spectacles and good theatre, trained a very effective army, built straight roads, protected shipping lanes, and expanded citizenship. Ironically, however, the historical evidence suggests that people living in Roman areas were shorter and had weaker bones than people before or after the Roman Empire—likely because the presence of disease impaired nutrition (Harper 2017). The ease of movement throughout the Empire brought not only better trade but the plague as well that killed far more Romans than the ever-ongoing battles to expand and sustain the Empire. Roman engineers did not successfully address sewage discharge and Roman medicine had not discovered the germ theory of disease. Great plagues would continue to bring sobering numbers of deaths through the middle ages. The aristocracy and the poor both died in great numbers, but as Boccaccio points out in the *Decameron* leaving plague infested areas may have helped the more privileged to survive (Tuchman 1978).

Movement remains an important force in the spread of disease today. The scale of human movement is considerable: from tourists and owners of second homes to traveling salesmen, concertgoers and attendees at international sporting tournaments, migrant laborers and refugees, hundreds of millions of people are on the road. Movement may mock borders as boundaries to restrict the spread of disease. This is especially true for diseases with longer incubation periods such as influenza (up to 4 days), measles (10–12 days), COVID-19 (up to 14 days), or Ebola (up to 21 days), as people may complete their travels before they realize that they are becoming ill.

Each year, seasonal influenza strains emerge from southern areas of Asia and spread across the globe, infecting an estimated billion people (WHO 2019). The predominant strains of seasonal influenza vary somewhat from year to year and health officials seek to predict them sufficiently in advance to enable manufacture of appropriate vaccines. Pandemic influenza occurs when a novel strain emerges to which a significant proportion of the population has no immunity. Fears are that a contemporary pandemic would spread even more quickly and be particularly lethal in areas of the world with inadequate access to health care for prevention or support (WHO 2019, p. 5). We are seeing these fears materialize today with COVID-19 spread.

The influenza virus is particularly unstable and subject to rapid mutation through genome re-assortment from various strains (Shao et al. 2017). Proximity can be harmful in exchanging different viral strains, as occurs in areas of Asia. Air transit moving people rapidly across the globe can then spread novel strains far afield.

To illustrate, the European Centre for Disease Prevention and Control estimates the scope of potential for influenza transmission by air, given that over 900 million airplane passengers travel within Europe every year: "The transmission of influenza viruses, for example, is facilitated in closed/semi-closed settings through direct person-to-person contact or from contaminated surfaces. At the beginning of the influenza A(H1N1) pandemic in 2009, air travel was the cause of the introduction of this new virus into countries not primarily affected, and aeroplanes are likely to be a major vector when the next pandemic occurs" (ECDC 2017). Indeed, disease transmission is a significant concern within the heart of the European Union with its relatively open land borders and free movement of goods and services within the Union. The EU may implement public health measures restricting travel on the part of people arriving by ship or plane from foreign destinations. At the height of the COVID-19 pandemic, borders were closed within the EU. But the scale of travel within the EU is of such a proportion that maintaining such a strict public health regime is challenging. The estimate is that 90 million intra-EU visits take place to France in any given year while another 80 million visits occur to Spain. Fortunately for the EU, the broad health of people living in Europe is superior to that in many parts of the globe, which suggests that the quality of personal health care systems should be seen in tandem with public health surveillance.

On a global scale, the World Health Organization tracks influenza and its spread. The WHO strategy for influenza during the decade from 2020 to 2030 follows the International Health Regulations (IHR) by seeking to build strong national capacities for influenza preparedness and response. Under the IHR, the state performs the primary and necessary role in the enterprise of legislation regarding health policy (WHO 2005, Art. 3 §4). The influenza strategy recognizes the primary role of states in developing health infrastructures and ensuring universal access to health care (WHO 2019, p. 6). By 2030, the desired outcome is for every country to have an evidence-based influenza plan, optimized to fit its needs (p. 9). This structure recognizes states as autonomous actors in governing their respective populations.

The IHR structure also commits states parties to obligations regarding the risk of disease spread beyond borders. Through becoming a state party to the WHO, states agree to developing and maintaining adequate surveillance capacities (WHO 2005, Art. 5) and to notifying WHO of events that may constitute public health emergencies of international concern (Art 6).

A major challenge to this regime is the problem of equity that we discussed in Chapter 4. Although states share the obligations to surveille and report threats equally, they may be far from equal in their commitment to collaborate with the WHO in addressing threats or in their ability to devote resources to surveillance. The IHR permit WHO to offer states collaboration in the effort to assess the potential for international spread through travel and, if states decline the offer of collaboration, to share the information with other states depending on the magnitude of the risk (Art. 10, §§3, 4). But the IHR do not give WHO authority to compel collaboration by states parties that are unwilling to collaborate. States also may be far from equal in their surveillance capabilities or their ability to address emergent health care needs. The IHR permit WHO to offer assistance to states in developing and maintaining their capacities (Art 5, §3). They also urge states to undertake to collaborate with each other, including through mobilization of financial resources, but this is a recommendation not a requirement (Art. 44).

An important provision in the IHR allows non-state actors to report possible health emergencies of international concern (Art 9). Receipt of information from non-state actors may lead WHO to start an investigation, even if the relevant state party has not notified WHO of concerns. Non-state actors may be engaged across borders in ways that states cannot be, along with other advantages and disadvantages. They may have greater capacities and access to information that states do not. Their inclusion may reflect concerns about willingness of states to cooperate with WHO. WHO has developed a framework of engagement with non-state actors in response to the recognition of the increasing role they are playing in global health (WHO 2016). However, participation of non-state actors in surveillance may have mixed results. WHO singles out concerns for transparency, undue influence, conflicts of interests, and potential risks to its credibility raised by partnerships with non-state actors. WHO is also concerned that through the support they are given non-state actors may acquire competitive advantages, pursue their own interests, and whitewash their images. The WHO Director-General reports annually on the status of engagement with non-state actors. The possibility of reporting may also be risky for non-state actors; Médecins Sans Frontières was forced to close treatment centers due to attacks during the 2019 Ebola outbreak in the Congo.

The ease by which influenza and other infectious diseases cross boundaries with mobile populations has led to increasing resistance to migrants such as asylum seekers and refugees. At this juncture, security policy and public health policy may seem to converge. Defending the city, the state and the nation against invading armies may become an obvious parallel with the image of an epidemic moving toward the city threatening illness and death. But engaging the enemy in a theater of war distant from the borders of a home state is not at all parallel to addressing contagious disease transmission. People coming from abroad may be seen as dangerous threats, but they are also desperate refugees. Demands to close borders may become insistent—regardless of their likely efficacy. On the other side, those living in areas where disease have emerged may face grim choices between isolation, flight, or the possible hope of inoculation or effective treatment.

States' obligations as parties to the IHR also involve expectations about how they are to respond to these perceived threats. These include limitations on the health documents that can be required of travelers who are not seeking temporary or permanent residence (Art. 35). They also include limitations on the health examinations of travelers that may be carried out without their informed consent (Art. 23, §3). Additional health measures are permitted on a case by case basis, provided that they are the least intrusive and invasive possible to achieve the objective of preventing the international spread of disease (Art. 23, §2). States are also permitted to collect information about destinations so that they may follow up if there is need to contact travelers (Art 23, §1).Travelers who refuse may be denied entry and states may take further action to prevent disease spread, including quarantine according to state law (Art 31, §2). In implementing any of these measures, the IHR require states to treat travelers in accord with respect for dignity, human rights, and fundamental freedoms (Art. 32). States must take gender, cultural, and ethnic or religious concerns into account and must arrange for food and water, appropriate accommodations and clothing, appropriate medical treatment, and protection of possessions (Art. 32). Summarized, these provisions allow states to protect public health but require them to minimize harm to individuals affected.

Reactions in the U.S. to Ebola illustrate the tensions that may affect how states actually perceive and respond to apparent threats from abroad. The West African Ebola outbreak in 2014–2016 presented no systemic threat to the U.S. homeland but was met with great fear and at least some stigmatization of people from the region as potential sources of infection. The only way Ebola could affect people within the U.S. who had not traveled to areas where they might be exposed was by coming into direct contact with infected bodily fluids from someone who had become infected abroad and returned to the US, yet misinformation about modes of Ebola transmission circulated widely. The failure to recognize a case of Ebola in a patient who had traveled from Liberia, and subsequent infection of health care workers in an underprepared facility, contributed to public perceptions that CDC had gravely underestimated the threat. A good deal of popular support was expressed by political leaders at local, state and federal levels for denying entry or re-entry of people who came from African nations experiencing the outbreak and for quarantine of people suspected of potential exposure. These calls extended to doctors and nurses who had served in charitable roles in countries associated with the Ebola outbreak. Health communication scholars Shaunak Sastry and Alessandro Lovari (2017) argue that CDC messaging about Ebola through social media reaffirmed western anxieties about infections emerging abroad. The image of President Obama deploying troops abroad to fight the epidemic may have contributed to this perception of foreign threat to be fought elsewhere. Other communication scholars note the prevalence of stories on social media about risks of infection in the U.S. in contrast to stories about the epidemic in West Africa or the science of Ebola spread (Roberts et al. 2017). A later critical review of the U.S. response to Ebola indicated the importance of involving local health departments in developing guidance, providing healthcare facilities and other potentially involved personnel such as EMS workers with appropriate guidance about infection control, communicating about risk to address popular misconceptions, and most importantly striving to avoid disconnections between these popular misconceptions and decisions made by political leaders (Dwyer et al. 2017).

In the wake of the West African Ebola epidemic, the CDC engaged in a rulemaking process to amend its authority to over quarantine of interstate travelers and travelers from abroad (HHS 2016). The proposed regulation expanded CDC's authority to monitor threats including through electronic means. It established requirements for commercial passenger flights to report death or illness to CDC and allowed CDC to implement travel restrictions on people moving among states who are reasonably believed to be infected with a quarantinable disease. Comments on the proposed rule were extensive and the final rule incorporated some protections in line with the IHR requirements for treatment of people under detention as potential threats to public health (HHS 2017). For example, under the final rule individuals being detained must be provided with adequate food and water, appropriate medical treatment and accommodation, and means of necessary communication. Individuals must also be informed that health examinations will be conducted by licensed health care workers with their informed consent. Critics argue, however, that the final rule is insufficiently protective in the extent to which it gives CDC discretionary authority to act quickly before administrative processes can occur (e.g. Edwards et al. 2017). In a notice of proposed rulemaking in July 2020, the administration of President Trump used pandemic threats as justification for emergency powers to reject asylum seekers to the U.S.

Nonetheless, people may be motivated to try to escape infection through movement, just as those who were able did in the Renaissance times of Boccaccio. Individuals behave rationally when they weigh the costs of leaving—moving away from families, jobs, or established networks—against the potential safety of flight. Movers may face the barriers of quarantine on arrival if domestic fears are sufficiently elevated. The IHR seek to mitigate the crueler aspects of this choice through protections for basic human dignity, and they may be well advised to do so. For diseases may travel very quickly with people who seek to hide concerns about their health. US reactions to the Ebola threat may prove instructive about the importance of protecting potential victims for achieving public health goals. One study indicates that of travelers to New York City being monitored for potential Ebola by the health department, a small but significant percentage gave misinformation about their temperatures or their whereabouts (Tate et al. 2017). Reported reasons included the stress of being monitored, discrimination at work, and avoidance by people they knew. Support during the monitoring, including information they could give to people about the protections monitoring could provide, was perceived as particularly helpful by people in complying with their surveillance. These findings support the need for addressing risks of harm to those under surveillance, when public health takes justified protective measures, to the ultimate success of protective measures, as Mark Rothstein (2015) argues the U.S. failed to do in reacting to Ebola.

In responding to movement, public health seeks to stabilize transmission by encouraging people to stay in place in times of a communicable disease crisis. To do so, it must calibrate responses carefully, providing accurate information about surveillance activities and what they are revealing, while not evoking irrational fear. Efforts to inform and to engage local entities, from health departments to members of multiple communities, are critical in countering flows of misinformation. Failures to respect people under the watching eyes of public health surveillance may backfire into sequestration or flight. The mobility of populations and diseases they carry underscore these ethical imperatives.

7.9 Summary

The impetus to insist on informed consent rests on important aspects of respect for individuals. Individuals should be able to understand what is happening to them and participate in decisions that affect them. Individuals should be treated as sources of their own good. These aspects of respect matter even when justifications for surveillance rely on non-paternalistic considerations such as prevention of harm to others or achievement of health equity. This chapter has explored how these aspects of respect can be achieved without reliance on individual consent when surveillance affects individuals, groups, or communities, and even when it crosses international borders.

References

Barnes, Clarissa G., Frederick L. Brancati, and Tiffany L. Gary. 2007. Mandatory Reporting of Noncommunicable Diseases: The Example of The New York City A1c Registry (NYCAR). *AMA Journal of Ethics* 9 (12): 827–831.

Berg, Jessica Wilen. 2012. All for One and One for All: Informed Consent and Public Health. *Houston Law Review* 50 (1): 1–40.

Bloomberg School of Public Health. 2019. What Is Public Health? https://www.jhsph.edu/about/what-is-public-health/. Accessed 31 July 2020.

Brennan, Jason. 2016. *Against Democracy*. Princeton: Princeton University Press.

Bresnick, Jennifer. 2017. How Do Population Health, Public Health, Community Health Differ? *Health IT Analytics* [online] (July 19). https://healthitanalytics.com/news/how-do-population-health-public-health-community-health-differ. Accessed 1 Aug 2020.

Buck v. Bell, 274 U.S. 200 (1927).

Bureau of Labor Statistics (BLS). 2016. Volunteering in the United States—2015. https://www.bls.gov/news.release/pdf/volun.pdf. Accessed 1 Aug 2020.

Cantor, Julie D. 2019. Mandatory Measles Vaccination in New York City—Reflections on a Bold Experiment. *New England Journal of Medicine* 381: 101–103.

Centers for Disease Control and Prevention (CDC). 2017. CDC and Indian Country Working Together. https://www.cdc.gov/chronicdisease/pdf/CDC-indian-country.pdf. Accessed 1 Aug 2020.

———. 2019. HIV and African-Americans. https://www.cdc.gov/hiv/group/racialethnic/africana-mericans/index.html. Accessed 1 Aug 2020.

Chamany, Shadi, Lynn D. Silver, Mary T. Bassett, Cynthia R. Driver, Diana K. Berger, Charlotte E. Neuhaus, Namrata Kumar, and Thomas R. Frieden. 2009. Tracking Diabetes: New York City's A1C Registry. *Milbank Quarterly* 87 (3): 547–570.

Chunara, Rumi, Lindsay Bouton, John W. Ayers, and John S. Brownstein. 2013. Assessing the Online Social Environment for Surveillance of Obesity Prevalence. *PLOS/ONE*. https://doi.org/10.1371/journal.pone.0061373. Accessed August 1, 2020.

Coughlin, Steven S., Selina A. Smith, and Maria E. Fernandez. 2017. *Handbook of Community-Based Participatory Research*. New York: Oxford University Press.

Dolgon, Corey, Tania D. Mitchell, and Timothy K. Eatman, eds. 2017. *The Cambridge Handbook of Service Learning and Community Engagement*. New York: Cambridge University Press.

Dwyer, Katherine Schemm, Heather Misner, Sara Chang, and Neyline Fajardo. 2017. An Interim Examination of the US Public Health Response to Ebola. *Health Security* 15 (5): 509–518.

Edwards, Kyle, Wendy Parmet, and Scott Burris. 2017. Why the C.D.C's Power to Quarantine Should Worry Us. *The New York Times* [online] (Jan 23). https://www.nytimes.com/2017/01/23/opinion/why-the-cdcs-power-to-quarantine-should-worry-us.html. Accessed 1 Aug 2020.

European Centre for Disease Prevention and Control (ECDC). 2017. Infectious Diseases on Aircraft. https://ecdc.europa.eu/en/all-topics-ztravellers-health/infectious-diseases-aircraft. Accessed 1 Aug 2020.

Faden, Ruth R., and Tom L. Beauchamp. 1986. *A History and Theory of Informed Consent.* New York: Oxford University Press.

Faden, Ruth R., and Sirine Shebaya. 2016. Public Health Ethics. *The Stanford Encyclopedia of Philosophy* (Winter 2016 Edition), Edward N. Zalta (ed.). https://plato.stanford.edu/archives/win2016/entries/publichealth-ethics/. Accessed 1 Aug 2020.

Fairchild, Amy. 2006. Diabetes and Disease Surveillance. *Science* 313 (5784): 175–176.

Fairchild, Amy L., Daniel Wolfe, James Keith Colgrove, and Ronald Bayer. 2007a. *Searching Eyes: Privacy, the State, and Disease Surveillance in America.* Berkeley: University of California Press.

Fairchild, Amy, Ronald Bayer, and James Colgrove. 2007b. Privacy and Public Health Surveillance: The Enduring Tension. *AMA Journal of Ethics* 9 (12): 838–841.

Flanigan, Jessica. 2013. Public Bioethics. *Public Health Ethics* 6 (2): 170–184.

———. 2017. *Pharmaceutical Freedom: Why Patients Have a Right to Self-Medicate.* New York: Oxford University Press.

Goldman, Janlori, Sydney Kinnear, Jeannie Chung, and David J. Rothman. 2008. New Your City's Initiatives on Diabetes and HIV/AIDS: Implications for Patient Care, Public Health, and Medical Professionalism. *American Journal of Public Health* 98: 807–813.

Goodman, Richard A., Rebecca Bunnell, and Samuel F. Posner. 2014. What Is "Community Health"? Examining the Meaning of an Evolving Field in Public Health. *Preventive Medicine* 67 (Suppl. 1): S58–S61.

Hall, Ingrid J., C. Ashani Johnson-Turbes, and Kymber N. Williams. 2010. The Potential of Black Radio to Disseminate Health Messages and Reduce Disparities. *CDC Preventing Chronic Disease: Public Health Research, Practice, and Policy* 7 (4): A87.

Harper, Kyle. 2017. *The Fate of Rome: Climate, Disease, & the End of an Empire.* Princeton: Princeton University Press.

Health and Human Services Department (HHS). 2016. Control of Communicable Diseases NPRM (August 15). *Federal Register* 81: 54229–54316.

———. 2017. Control of Communicable Diseases Final Rule. *Federal Register* 82: 6890–6978.

Hertzman, Clyde, and Arjumand Siddiqi. 2009. Population Health and the Dynamics of Collective Development. In *Successful Societies: How Institutions and Culture Affect Health*, ed. Peter A. Hall and Michèle Lamont, 23–53. New York: Cambridge University Press.

HIV.gov. 2019. New Factsheets: HIV's Impact in the African American Community (Feb 22). https://www.hiv.gov/blog/new-factsheets-hiv-s-impact-african-american-community. Accessed 1 Aug 2020.

Israel, Barbara A., Eugenia Eng, Amy J. Schulz, and Edith A. Parker, eds. 2013. *Methods for Community-Based Participatory Research for Health.* 2nd ed. San Francisco: Wiley.

Jacobson v. Massachusetts, 197 U.S. 11 (1905).

Jewkes, Rachel, and Anne Murcott. 1996. Meanings of Community. *Social Science & Medicine* 43 (4): 555–563.

Kim, Nancy S. 2019. *Consentability: Consent and Its Limits.* Cambridge: Cambridge University Press.

Krent, Harold J., Nicholas Gingo, Monica Kapp, Rachel Moran, Mary Neal, Meghan Paulas, Puneet Sarna, and Sarah Suma. 2008. Whose Business Is Your Pancreas? Potential Privacy Problems in New York City's Mandatory Diabetes Registry. *Annals of Health Law* 17: 1–33.

Kymlicka, Will. 1996. *Multicultural Citizenship: A Liberal Theory of Minority Rights.* New York: Oxford University Press.

LaMotte, Sandee. 2019. Why New York Hasn't Contained the Largest and Longest Measles Outbreak in Decades. *CNN* [online] (Mar 29). https://www.cnn.com/2019/03/29/health/measles-ny-outbreak-fear-misinformation/index.html. Accessed 1 Aug 2020.

Lee, Lisa M., Charles Heilig, and Angela White. 2012. Ethical Justification for Conducting Public Health Surveillance Without Patient Consent. *American Journal of Public Health* 102 (1): 38–44.

Lukowski, Jerzy. 2012. "Machines of Government": Replacing the Liberum Veto in the Eighteenth-Century Polish-Lithuanian Commonwealth. *The Slavonic and East European Review* 90 (1): 65–97.

MacQueen, Kathleen M., Eleanor McLellan, David S. Metzger, Susan Kegeles, Ronald P. Strauss, Roseanne Scotti, Lynn Blanchard, and Robert T. Trotter II. 2001. What Is Community? An Evidence-Based Definition for Participatory Public Health. *American Journal of Public Health* 91 (12): 1929–1938.

Manson, Neil C., and Onora O'Neill. 2007. *Rethinking Informed Consent in Bioethics*. Cambridge: Cambridge University Press.

Mikesell, Lisa, Elizabeth Bromley, and Dmitry Khodyakov. 2013. Ethical Community-Engaged Research: A Literature Review. *American Journal of Public Health* 103 (12): e7–e14.

Morgan, Ethan, Britt Skaathun, and John A. Schneider. 2018. Sexual, Social, and Genetic Network Overlap: A Socio-Molecular Approach Toward Public Health Intervention of HIV. *American Journal of Public Health* 108 (11): 1528–1534.

O'Neill, Onora. 2003. Some Limits of Informed Consent. *Journal of Medical Ethics* 29: 4–7.

Orthodox Jewish Nurses Association (OJNA). 2019. About. https://jewishnurses.org/. Accessed 1 Aug 2020.

Parés, Marc, Sonia Ospina, and Cliff Frasier. 2017. *Social Innovation and Democratic Leadership: Communities and Social Change from Below*. Cheltenham: Edward Elgar.

Public Health Leadership Society. 2002. Principles of the Ethical Practice of Public Health. https://www.apha.org/-/media/files/pdf/membergroups/ethics/ethics_brochure.ashx. Accessed 1 Aug 2020.

Putnam, Robert. 2000. *Bowling Alone*. New York: Simon & Schuster.

Rajan, Raghuram. 2019. *The Third Pillar: How Markets and the State Leave the Community Behind*. New York: Penguin Press.

Roberts, Hal, Brittany Seymour, Sands Alden Fish II, Emily Robinson, and Ethan Zuckerman. 2017. Digital Health Communication and Global Public Influence: A Study of the Ebola Epidemic. *Journal of Health Communication* 22 (sup 1): 51–58.

Rothman, Joshua. 2016. When Bigotry Paraded Through the Streets. *The Atlantic* [online] (Dec 4). https://www.theatlantic.com/politics/archive/2016/12/second-klan/509468/. Accessed 1 Aug 2020.

Rothstein, Mark. 2015. Ebola, Quarantine, and the Law. *Hastings Center Report* 45 (1): 5–6.

Sastry, Shaunak, and Alessandro Lovari. 2017. Communicating the Ontological Narrative of Ebola: An Emerging Disease in the Time of "Epidemic 2.0". *Health Communication* 32 (3): 329–338.

Schloendorff v. Society of New York Hospital, 105 NE. 92 (N.Y. 1914).

Shao, Wenhan, Xinxin Li, Mohsan Ullah Goraya, Song Want, and Ji-Long Chen. 2017. Evolution of Influenza A Virus by Mutation and Re-Assortment. *International Journal of Molecular Sciences* 18 (8): 1650.

Shore, Nancy, Ruta Brazauskas, Elaine Drew, Kristine A. Wong, Lisa Moy, Andrea Corage Baden, Kirsten Cyr, Jocelyn Ulevicus, and Sarena D. Seifer. 2011. Understanding Community-Based Processes for Research Ethics Review: A National Study. *American Journal of Public Health* 101 (S1): S359–S364.

Silverman, Ross D. 2019. Controlling Measles Through Politics and Policy. *Hastings Center Report* 49 (3): 8–9.

Stoto, Michael A., Cynthia Abel, and Anne Dievler, eds. 1996. *Healthy Communities: New Partnerships for the Future of Public Health*. Washington, DC: National Academies Press.

Sullivan, Patrick S., David W. Purcell, Jeremy A. Grey, Kyle T. Bernstein, Thomas L. Gift, Taylor A. Wimbly, Eric Hall, and Eli S. Rosenberg. 2018. Patterns of Racial/Ethnic Disparities and Prevalence in HIV and Syphilis Diagnoses Among Men Who Have Sex With Men, 2016: A Novel Data Visualization. *American Journal of Public Health* 108 (S4): S266–S273.

Tate, Anna, Ifeoma Ezeoke, David E. Lucero, Chaorui C. Huang, Alhaji Saffa, Jay K. Varma, and Neil M. Vora. 2017. Reporting of False Data During Ebola Virus Disease Active Monitoring-New York City, January 1, 2015–December 29, 2015. *Health Security* 15 (5): 509–518.

Tocqueville, Alexis de. 2006 ed. *Democracy in America.* Translated by Harry Reeve. Project Gutenberg. http://www.gutenberg.org/files/815/815-h/815-h.htm. Accessed 1 Aug 2020.

Tuchman, Barbara. 1978. *A Distant Mirror: The Calamitous 14th Century.* New York: Alfred A. Knopf.

United Nations (UN). 1967. International Covenant on Economic, Social and Cultural Rights. https://treaties.un.org/doc/Treaties/1976/01/19760103%2009-57%20PM/Ch_IV_03.pdf. Accessed 1 Aug 2020.

———. 2007. United Nations Declaration on the Rights of Indigenous Peoples. https://www.un.org/development/desa/indigenouspeoples/wp-content/uploads/sites/19/2018/11/UNDRIP_E_web.pdf. Accessed 1 Aug 2020.

Von Gierke, Otto. 1939. *The Development of Political Theory.* Translated by Bernard Freyd. New York: W.W. Norton Co.

Weinstein, James N., Amy Geller, Yamrot Negussie, and Alina Baciu, eds. 2017. *Communities in Action: Pathways to Health Equity.* Washington, DC: National Academies Press.

World Health Organization (WHO). 2005. *International Health Regulations*, 3rd ed. https://apps.who.int/iris/bitstream/handle/10665/246107/9789241580496-eng.pdf?sequence=1. Accessed 1 Aug 2020.

———. 2016. Framework of Engagement with Non-State Actors (May 28). http://apps.who.int/gb/ebwha/pdf_files/wha69/a69_r10-en.pdf. Accessed 1 Aug 2020.

———. 2019. Global Influenza Strategy 2019–2030. https://apps.who.int/iris/bitstream/handle/10665/311184/9789241515320-eng.pdf?ua=1. Accessed 1 Aug 2020.

World Health Organization (WHO) director-general. 2019. Engagement with Non-State Actors: Report by the Director General (Nov 23). http://apps.who.int/gb/ebwha/pdf_files/EB144/B144_36-en.pdf. Accessed 22 June 2019.

Wuthnow, Robert. 2013. *Small-Town America: Finding Community, Shaping the Future.* Princeton: Princeton University Press.

Chapter 8
Conclusion

In concluding, we return to the ubiquity of surveillance by quoting from three of the major public health entities responsible for surveillance today:

> From the World Health Organization: "Public health surveillance is the continuous, systematic collection, analysis and interpretation of health-related data needed for the planning, implementation, and evaluation of public health practice."
>
> From the European Commission: "Surveillance systems provide information for monitoring communicable disease trends and early detection of outbreaks, helping to identify risk factors, and areas for intervention. They provide information for priority setting, planning, implementation and resource allocation for preventive programmes and for evaluating preventive programmes and control measures.
>
> And from the U.S. Centers for Disease Control and Prevention: "Public health surveillance is the continuous, systematic collection, analysis and interpretation of health-related data needed for the planning, implementation, and evaluation of public health practice. The Center for Disease Control"

Surveillance in the public health context is continuous. It is inclusive of much of what the WHO means by health as the achievement of the highest attainable standard of wellbeing, not merely the absence of disease, and certainly not merely the absence of communicable disease. These mission statements make clear the extent of surveillance's reach.

Throughout the volume, our argument has been that although the strength of the commitment to surveillance is undiminished, how we gather data continues to change and often dramatically so. Surveillance has moved far beyond counting numbers in the population to tracking movements, sensing bodily change, and observing how people do and might be brought to communicate with each other through social media. Public health, and perhaps even more so other actors entering the surveillance enterprise have access to heretofore unimagined uses of technology within and outside the body, the home, the work environment, and the public square. Building and sustaining confidence in technology that is fluid and changing will come to matter more not less.

© Springer Nature Switzerland AG 2021
J. G. Francis, L. P. Francis, *Sustaining Surveillance: The Importance of
Information for Public Health*, Public Health Ethics Analysis 6,
https://doi.org/10.1007/978-3-030-63928-0_8

Sustaining surveillance, we have argued, will require multifaceted efforts in different circumstances. Insistence on ethical science is key, as is concern for those under surveillance as themselves potential subjects of illness in need of care. Equity in surveillance is critical; subpopulations must not be ignored, and some must not be used as mere sources of information for others. When surveillance reaches beyond contagion and harm to others, care must be taken to respect the values of individuals and groups. As the WHO Guidelines discuss well, transparency, engagement, and respect for cultural differences are needed. Surveillance has great potential to improve the health of individuals, groups, communities, countries, and populations, but only when it is conducted in ways that can generate warranted trust.

CPSIA information can be obtained
at www.ICGtesting.com
Printed in the USA
LVHW082022080123
736723LV00006B/361